THE Invisible Kingdom

From the Tips of Our Fingers to the Tops of Our Trash, Inside the Curious World of Microbes

Idan Ben-Barak

A MEMBER OF THE PERSEUS BOOKS GROUP
New York

Published in 2009 by
Basic Books, A Member of the Perseus Books Group
387 Park Avenue South
New York, NY 10016-8810

Published in 2008 in Australia by Scribe Publications as *Small Wonders: How Microbes Rule Our World*

Designed by Timm Bryson

Library of Congress Cataloging-in-Publication Data
Ben-Barak, Idan.
The invisible kingdom : from the tips of our fingers to the tops of our trash, inside the curious world of microbes / Idan Ben-Barak.
p. cm.
"Published in 2008 in Australia by Scribe Publications as Small Wonders: How Microbes Rule Our World."
Includes bibliographical references.
ISBN 978-0-465-01887-1 (alk. paper)
1. Microbiology—Popular works. I. Title.
QR56.B38 2009
579—dc22
2009019655

ISBN-13: 978-0-465-01887-1

10 9 8 7 6 5 4 3 2 1

For Tamar, for life

And now we rise
And we are everywhere

NICK DRAKE, "From the Morning"

CONTENTS

PROLOGUE

Q: Hello, hello, hello, what's all this, then?

A: A book about microbes.

Q: Come again?

A: Microbes. Microorganisms. Germs. Bugs. Very small living things.

Q: Oh, God, is that the time? Sorry, must rush . . .

A: Oh no you don't. . . . Now stop squirming, will you? It's quite a good book.

Q: It's full of science, isn't it? With graphs and whatnot?

A: No graphs.

Q: How about charts?

A: Nope.

Q: Equations?

A: No. Oh, sorry, there is one, actually. It's 1 + 1. And I got the answer wrong.

Q: Footnotes?

A: Dozens. I can't get rid of the blasted things. They're like textual parasites, popping up all over the place. Anyway, they're not the usual sort. You'll see.

Q: Any numbers at all?

A: Here and there. A few largish ones lurk at the end of Chapter 1. But don't worry; they're well trained.

Q: It's a book about germs, you say?

A: Yes.

Q: I saw this film where . . .

A: Wait. Don't tell me—the human race is nearly wiped out by a genetically engineered virus that escapes from the lab and either kills everyone or turns them into zombies? I'm still waiting for a film in which the virus is tall, brave, handsome, and witty and gets the girl at the end. That would be something to watch. In any case, both scenarios are about equally realistic.

Q: But is the book all about a whole lot of horrible diseases?

A: Not at all. Diseases get an honorable mention, of course, but I've tried to leave sufficient room for other important stuff, like sex, burping sheep, politics, slimy gunk, rocket fuel, genes, futuristics, and computers. Also, for some reason, frogs. They're constantly underfoot.

Q: Are there any helpful health tips inside?

A: One or two. More if you're a cow.

Q: Any useful information at all?

A: There's some bad investment advice in Chapter 3, if that helps any.

Q: So why should I read it?

A: Because otherwise you'll never know.

Q: Right. What happens now?

A: I untie you, and we begin. We start off with one of the most puzzling, inexplicable, and complex phenomena known to humankind.

Q: The origin of life?

A: Major-league baseball.

CHAPTER 1

Bugs on Display

Here is an excerpt from a *New York Times* sports section of June 2008: "The Twins scratched a run in the second, then tied it in the third thanks to an error by Pettitte, who threw the ball into the Twins' bullpen trying to pick off Carlos Gómez at first base. Gómez scored on Alexi Casilla's safety-squeeze bunt single."

Unfortunately for me, I don't know much about baseball, so I don't have a clear idea of what's going on. How do you scratch runs? Does it require special equipment? What's a squeeze bunt? Who's squeezing whose bunt? Is it considered a naughty thing to do? What are twins doing in a bullpen? How many bulls, if any, are in there with them?

It's no use. If I want to appreciate baseball, I have to sit down with someone who'll explain what's going on, first. It's the same with just about anything from hot rods to quantum mechanics: If you really want to get it, you need to speak a bit of the language and understand a few of the rules; otherwise, it just seems like a whole lot of running around.[1]

Understanding jargon isn't a mark of intelligence or ability; it's just a matter of becoming familiar with the subject. Case in point:

1. Or, in the case of baseball, a lot of standing around.

My favorite pastime in a supermarket line is to try to figure out what the tabloid headlines mean, without peeking inside.[2] Not Nobel Prize–winning stuff, you'd think, but it can be quite a puzzle if I haven't been keeping track of current events in the celebrity sector.

The jargon barrier is a simple-enough principle, but we tend to forget it. Professionals of every kind use special terminology that sounds very impressive to the outsider, but is usually nothing more than shorthand for something that could be understood by anyone, given a little time and a dictionary.

I want to tell you a few stories about microbes, but I have a problem: If I go into detailed and rigorous explanations of biological terms and ideas, it would take a lot of time and paper, this book would become a textbook, and I'd lose you. On the other hand, if I just start yapping on about sigma factors and siRNA, you might decide to tell me to get lost.

I don't want to turn you into a microbiologist; being a microbiologist is what microbiologists were put on this Earth for. So I've opted for the middle path: a quick run-through of some of the basics, names, and principles of biology. If you want to know more, have a look at the Further Reading section at the back of the book and, if you get lost along the way, be sure to refer to the glossary.

What Is a Microbe, Anyway?

Almost everyone knows the cycle of life by heart: Plants get energy from the Sun and nutrients from the soil, animals eat plants and each other and then die, microbes break down dead animals and plants into nutrients, and then it starts again.

---

2. I'm also secretly waiting for this headline: ANGELINA SEEKS SHORT, SCIENCE-WRITING LOVER TO MEND BROKEN HEART.

But what exactly is a microbe? A microbe is a general name for any creature that is, individually, too small to be seen with the unaided eye. This definition is very old and very loose, so it embraces a lot of different sorts of creatures: bacteria (the group we commonly think of when we say "germ"), archaea (superficially resembling bacteria but recently found to be quite different in many respects), fungi (from yeast to mushrooms), and protists (this group includes primitive algae, amoebas, slime molds, and protozoa). Viruses are microbes, too, but we'll save for later the juicy question of whether they're truly alive or not.

These groups are as different from one another as we are from them—usually even more. From a microbe's point of view, you are virtually identical to, say, a flea because you and the flea share many processes and structures that the microbe doesn't. This is why an antibiotic can kill bacteria, but not people (or fleas): It jams a process that is unique to bacteria. This is also why antibiotics don't work on viruses, which aren't remotely like bacteria (which means that taking penicillin for your flu is useless); nor do they work on fungi, which have their own distinctive way of doing things, and which we need to develop special antifungal drugs to deal with.

A microbe is a single-celled creature. You and the flea are composed of many different types of cells that hold themselves together and depend on each other for survival: Your brain cells are useless without your liver, muscle, and heart cells, for example. A single microbial cell, however, is an independent creature that can survive and reproduce without help from other cells.[3]

Microbes are also rather teensy. Your average *Escherichia coli* (*E. coli*) cell is about two micrometers long, which means that it would

3. It's actually not that simple. We'll see later on how the plot thickens, but it's a good starting point.

take about 50,000 cells back to back (and a lot of convincing) to circle your little finger.[4] A typical virus is ten to a thousand times smaller than that, which means, proportionally, if a virus were the size of a tennis ball, you'd be big enough to lie down with your feet in San Diego and your head crushing the Golden Gate Bridge.

There's much yet to learn about microbes, including how prevalent they are, how intimately we are involved with them, and how much we rely on them for sustaining life on Earth.

I'll try not to hammer on too much about how grateful we should be to microbes for our continued survival. (Well, why should we? It's not as if they're doing it out of the goodness of their hearts, which they haven't got.) To be evenhanded, I'll also try to minimize any unnecessary preoccupation with disease and death. While I won't hesitate to delve into gruesome tales (some of which you'll dearly wish were fiction), it seems to me that the numerous harmful interactions between microbes and humans have received enough attention as it is; nonetheless, if you particularly relish matters of doom and gloom, refer to the Further Reading section for a few excellent books on that sort of thing.

Are you still lying there between San Diego and San Francisco? Come on, get up. We've got work to do. We need to take a bit of a hop backwards to look at the basic makeup of all living things, microbes included.

What You'll Wish I Hadn't Told You About DNA

Life on Earth began several billion years ago. We don't know exactly how, but current understanding is that it started with a molecule

4. 1 meter = 1 million micrometers. 1 foot = 304,800 micrometers.

floating in the ocean.[5] There were many kinds of molecules floating around, but this one was the first to have a special quality: It could collect material from its surroundings and use it to make a copy of itself. This copy would then make other copies, and so on. Pretty soon, there were a lot of these copiers around, copying away. Because they weren't perfect copiers, variations would sometimes appear in the new copy. If a random change occurred to make the molecule copy itself more quickly or efficiently, its copies would spread faster than the others.[6]

We don't know what the original copiers looked like, but it seems like a safe bet that they were similar to a type of molecule we call RNA (ribonucleic acid). This molecule, together with a very similar type of molecule called DNA (deoxyribonucleic acid), is responsible for very nearly all the things we refer to as living.[7]

DNA, DNA . . . we hear about it so often: Its famed double-helix structure has become a familiar design motif, and its name is now tossed around in everyday conversation. So why all the fuss?

Simply, it's because DNA is the cornerstone for every living thing: All organisms are made up of either one or many cells, and each cell contains DNA molecules, which comprise the cell's repository of information. If you're wondering, pure DNA looks and feels like cloudy

5. A molecule is a group of two or more atoms held in a stable configuration by chemical bonds. This definition begs the questions "what are chemical bonds?" and "what is an atom?" which leads us to protons, electrons, and neutrons, then electromagnetic forces, then quarks. . . . Where do we stop explaining?

6. The origin-of-life question is actually far from being that simple, but going through it in any detail will take me a couple of chapters, and we don't want that. In addition to the science, the whole issue has somehow become a hotly contested political topic nowadays, with various parties insisting that this couldn't possibly have happened all by itself and that there was surely someone else involved—suspects include the CIA, Greenpeace, and the Freemasons.

7. With the exception of prions, which I will tell you about in Chapter 8.

white snot. You'd think this wondrous material should be some sort of golden, shimmering filament, but nature doesn't give a hoot about our aesthetic sensibilities.

When we refer to DNA, we usually talk about its *sequence*. Just as low-level computer code is made up of a string of ones and zeros, a DNA molecule encodes information using a genetic code that is made up of a string of four alternative bases: A, T, G, and C (adenine, thymine, guanine, and cytosine). A DNA sequence looks something like this, when written down on paper: . . . ATTTGCAGTT-TACCCGTG . . . To us, this is as meaningless as binary code, 000101010000100, but the cell's mechanisms know how to read it.

The total genetic information encoded in these sequences is called the genome. It's important to understand that there are a lot of different kinds of information in a genome. There are bits that tell other bits how and when to work, bits that point to other bits, and bits whose function, if there is one, we don't know yet. The most straightforward bits, however, are called genes.

Gene: another popular word that's bandied about all over the place. Mercifully, we encounter far fewer misleading ideas about genes in the media nowadays (for example, the notion that you can have "a gene for" complex traits like aggressiveness, depression, or fashion sense), but there's still a lot of creative misunderstanding of the concept. Part of the problem is that even biologists have differing definitions of the term. The most popular include "a hereditary unit," which is useful for theoretical evolutionary studies, and the more hands-on explanation, which lab people favor: "a DNA sequence responsible for making a functional product." No definition is the correct one, because people who didn't know, and couldn't possibly have known, what DNA was (or how it worked) coined the term way back in the nineteenth century. Sixty years later, the theoretical concept suddenly became a physical reality—something you could actually see and touch (if you enjoy handling snot, of course).

So a microbe is a single cell, a cell contains a genome, and genomes are made up of genes (and some other stuff). Now for the part you've all been waiting for: reproduction.

DNA is packed as two complementary strands in a cell. This enables the cell to create two copies of its genome when it divides, so that each new copy of the cell (the "daughter cell") gets a complete copy of the information it needs to function.

In a bacterial or archaeal cell, the bulk of the DNA is contained in one circular molecule called a chromosome. In fungi, protists, and most multicellular organisms, we find more than one chromosome. A human cell contains twenty-three different kinds of chromosomes with two copies of each kind, because humans reproduce sexually, and sexual reproduction is all about the offspring receiving one copy of each chromosome from each parent. Microbes, on the other hand, reproduce asexually: The microbial cell makes a copy of its genome, divides itself in two (so that each half receives one whole copy), and both halves start growing again. Cycle completed. No arguing about which parent the kid looks like, or whose nose she got.

The Music of the Genes

Genes, by themselves, don't *do* anything; rather, they're instructions for doing things. A genome floats around by itself, doing nothing, until another component of the cell (an enzyme called RNA polymerase) grabs it and "reads" a certain gene off it, running over that particular stretch of DNA to produce a copy, or transcript, of that gene (or multiple copies, in some cases). It's a bit like what would happen if you took a lengthy instruction book and photocopied just the one page relevant to the specific product you needed; if you needed many workers to use the page at once, you might then make multiple copies.

An intricate apparatus called a ribosome then uses the RNA transcript to create a protein molecule.[8] Which protein molecule? It depends on the gene that's being read.

Proteins are a very large category of molecule. There are countless types of proteins, and they do an astounding variety of things. If a cell were a house needing to be built, proteins would be the tools, the craftsmen, and the building materials.

The "building materials" are called structural proteins (good examples of these are the actin and tubulin fibers that prevent our cells from collapsing into themselves, and keratin, which makes up our hair and nails), and the "craftsmen and tools" are called enzymes. Every type of enzyme is able to do one specific thing: They attach themselves to a specific target molecule, or substrate, and then act on it in some way, either by cutting it, joining it, or modifying it.[9] Digest food? Enzymes. Move a muscle? Enzymes. Think? Enzymes.

Another category of protein moves stuff around. They latch on to a target molecule and then either move the molecule from place to place (transport proteins), or just signal to the cell that they have caught and identified a particular molecule (receptors).

Proteins can also act as the signals themselves—hormones (for example, insulin) are a well-known class of these signal proteins.[10]

8. This is a ridiculously short description of a process that takes a student many long, caffeine-enriched nights to memorize, and that took scientists several decades to work out.

9. A substrate is any material that an enzyme works on.

10. Many hormones, though, such as adrenaline or testosterone, are not proteins.

It's very important at this point to emphasize that a cell doesn't just read its entire genome from start to finish and churn out all the proteins that the genes encode. Nearly every cell in your body has a complete set of genes, but they don't use them all. Your skin cells don't produce liver enzymes, and your liver cells don't try to become muscles. There are very elegant regulatory mechanisms in place to make sure that each cell expresses only those proteins (and only the correct amounts of each protein) that it should. If all proteins were expressed at once, it would be a chaotic jumble—like the meaningless din that you create (rather than music) when you hit all the keys on a piano at the same time.

Remember the oft-quoted scientific fact that says that humans share 98 percent of their genes with chimpanzees? Although it seems to suggest how alike the two species are, that's not quite the case. True, we share a lot of genetic information with our chimpy cousins, but it's read and used in very different ways (the growth of body hair springs to mind). As usual, it's not what you have that really matters: It's what you do with it.

The microbial cell works much the same way. Because it's single-celled and does everything by itself, it does use its genome quite comprehensively—but not all at once. It can sense its surroundings and react to them, produce enzymes that break down a particular nutrient when that nutrient is around, and switch to different enzymes if there's another, better nutrient to be used. If it's hot, it manufactures proteins that help cope with heat (disappointingly, they don't look like tiny electric fans), and it disassembles and recycles these when it cools down again.

If we combine all this together, we can visualize what a living cell looks like: There is a genome (a very long DNA molecule), which is being manipulated by regulatory proteins and RNA polymerase in order to churn out shortish RNA sequences. These sequences go out to the ribosomes, which then make all sorts of new proteins. Around the genome, thousands of processes are going on—enzymes are working on substrates, proteins are moving material from place to place, and things are constantly being built up, broken down, replaced, and modified. If you picture a factory reduced by a scale of about 100 million, sped up ten-thousandfold, duplicating itself occasionally, you'll get the general idea.

All this activity is contained within a cellular membrane, which separates "inside" from "outside."[11] On (and in) the membrane are receptor and transport proteins, which are the cell's link to the outside environment: They sense it, react to it, and bring things in and out of it.

In humans, this environment is mainly composed of neighboring cells: As I mentioned before, a human cell is a machine inside a machine—it can only survive and function as a component of a much larger structure.

A microbe's environment, on the other hand, may contain its siblings (read: competitors), members of other microbial species (read: competitors), and things that eat microbes (read: bad news). Important physical conditions—temperature, salinity, energy sources, the chemical composition of its surroundings—also affect the microbe, every second of its life. A microbe attempting to set up house inside a human lung, for example, has to deal with our immune system trying to get rid of it, while a microbe in a muddy puddle, or up a tree, has a very different set of problems.

---

11. Interestingly, some pretty convincing theories say that membrane formation, not molecule replication, was the crucial step in the origination of life. Or perhaps it was both of those together. Or replication *of* a membrane . . . as I said, it's not simple.

Reproduce. Repeat. Repeat.

We know now what a microbe is, what it looks like, and a little bit about how it works; but what does it *do*?

Mostly, it multiplies. A microbial cell is an object primarily concerned with getting enough energy and materials to repeatedly duplicate itself. This is not because it wants to (microbes don't have brains to want things with), but because that's the basic principle of evolution—if something tends to reproduce effectively, it will spread. If it doesn't, its numbers will dwindle away into nonexistence so that, eventually, we just don't see microbes that aren't that good at multiplying around anymore.

Therefore, microbes that have randomly developed something to help them survive and multiply in their environment will (nonrandomly) outbreed the others. That is the classical theory, at least; but, later on, I'll introduce you to some microbes that also develop tricks and shortcuts to speed up and optimize their evolution.

On the flip side of the race, some microbes actually lose something to gain an advantage, rather than developing something. As professionals in any field of racing will tell you, stripping the machine down is absolutely crucial for gaining a competitive edge because—and this is a principle that is as important in biology as it is everywhere—everything has its price. If a microbe has lost an unnecessary system, it no longer needs to allocate time, material, and energy towards keeping that system functioning, and it can put more resources into reproduction. That microbe is one step ahead in the perpetual race. Sometimes, a particular system or quality isn't necessary for a microbe's survival; for instance, the ability to survive intense heat is useless in a cool environment, so after some time, most microbes in a cool environment would be the offspring of the first individual microbe to have lost that ability.

Of course, not all environments are the same. After the first cells came into being, the natural optimization process meant that if you saw the same type of microbe in two different environments, over time, each would develop in radically different ways.[12] Microbes living in warm places, for instance, became better and better at dealing with high temperatures, and could venture out further into hotter places. Why would they adapt like that? Because of the reward: A microbe that could move out into new territory would have the place to itself. We're not just talking about physical territory here: If a microbe evolves so that it is able to use something in its present environment in a beneficial way (say, a food resource), it will suddenly be much better off, and will be able to reproduce the hell out of everyone else.

And reproduce they do: Given good conditions, an *E. coli* cell can become two cells within twenty minutes. Multiply this process by a lot, give it a few billion years to run, and you get what we see today: Microbes of all sorts, everywhere, swimming in every droplet of water, and hanging on to every surface.

---

12. If you existed a few billion years ago and had the patience to hang around in boiling mud for years and years.

BONUS TRACK #1

Doing the Numbers

Here are a few impressive figures about microbes, specifically designed to knock your socks off:

- Number of microbes in teaspoonful of garden soil: about a trillion
- Number of species of microbes in that teaspoonful of garden soil: about 10,000
- Total number of microbial species: nobody knows; somewhere in the lower billions is the best estimate
- Number of microbes per square centimeter of human skin: upwards of 100,000
- Ratio of microbial cells to human cells in the human body: 10 to 1
- Overall weight of microbes in a healthy human body: 2 to 4 pounds.
- Number of times an *E. coli* cell is able to reproduce in a day: 72
- Period in which microbes first appeared on Earth: 3.8 billion years ago
- Period during which they had the place to themselves: about 3 billion years
- Period in which the human race first noticed that microbes existed: the seventeenth century
- Period in which the human race started noticing that microbes were causing disease: the late nineteenth century

These numbers don't even take viruses into account; if they did, the figures would escalate by whole orders of magnitude.

CHAPTER 2

Bugs on the Map

I've always wanted to see catchier naming schemes for microbes. Where are the aye-aye, the okapi, the bumpy rocketfrog, the turbo snail, and the bandicoot of the microbial world? The coolest-sounding microbe I know of goes by the name *Actinomyces funkei*. Sure, *Dokdonella fugitiva* holds some dark, mysterious allure, *Albidovulum inexpectatum* maintains an element of surprise, and *Aeromonas popoffii* is to be commended for its airy jollity; but microbe names tend to be long, Latin, descriptive, and—to the layman—obscure (*Geodermatophilus obscurus* scores high marks on all counts). The only redeeming feature about microbial names is the exotic locations that they often hint at: *Actinoplanes brasiliensis*, *Desulfomicrobium norvegicum*, *Klebsiella singaporensis*, *Desulfovibrio mexicanus*, local candidate *Anaerobranca californiensis*, and the very specific *Dyadobacter beijingensis* are but a small sample of this category.

The name of a microbe usually marks where that particular species was first found. The dreaded Ebola virus, for example, was named after the river valley near where the first outbreak of the disease occurred. Geographical location, however, is hardly the most engaging thing about the places in which microbes find themselves. Microbes excel at surviving in unusual, even extreme, environments that are usually much too rough for other types of life.

Finding some of these microbes involves journeying to the ends of the Earth; but, before we do that, let's first stop at a slightly less captivating location to meet a microbe that's being grown by the billions—and find out why.

King of the Lab

Allow me to introduce you to *E. coli*, humble resident of our lower intestines that is grown relentlessly and studied in countless labs all over the world, and that causes nauseating smells in toilets everywhere.

There is, of course, no such thing as a typical microbe, just as there's no such thing as a typical human being; but *E. coli* will do to begin with, and it will give us something against which to compare some of the more extreme examples of microbehood.

E. coli belongs to the *Enterobacteriaceae* family of bacteria. Weighing in at one-millionth of a millionth of a gram (one gram is about the weight of a dollar bill), it measures a delightful two micrometers in length and about 0.8 of a micrometer in diameter, lending it a dashing, elegant, rodlike shape—not unlike a small, wriggly licorice bullet. Its hobbies include swimming around and reproducing itself. It usually hails from the lower intestines of mammals (a home that it shares with many other microbial species), but it has been found in many places, usually as a result of being excreted from the intestines into water. As befits its station in life, it prefers the sort of conditions that are found in its original habitat: wet, 98.6° F, alkaline, and rich in the nutrients that it lives off, and at the same time, it helps our digestive systems to process and absorb.

At first glance, there is nothing in particular about this microbe that seems to merit all this attention. True, it is in close personal contact with humans and it assists our digestive systems, but it hardly ever causes any trouble (there are a few strains of it that have gone bad, like the moderately dreaded and evocatively named O157:H7, but those strains make up a very small minority of its numbers) and it has no particularly captivating features. And yet biologists everywhere grow it, study it, and mess around with it constantly. Why is that? What's so special about it?

The answer is convenience—*E. coli* is the perfect laboratory tool. Several properties endear it to us so:

First, it can be grown quickly and cheaply. A single bacterial cell can be left to grow overnight inside a beaker that's placed in an incubator; the next morning, there will be billions of exact copies swirling around in there for us to study at our leisure.

Second, you can store it practically forever in a freezer, then thaw it, and carry on using it as if nothing happened. Try doing that with a lab rat.[1]

Third, it is quite accommodating: It accepts foreign DNA relatively easily. If you have a certain gene that you wish to study—never mind if it's bacterial, human, or any other sort—you can insert a single molecule of it into an *E. coli* cell, chuck the microbe into a beakerful of broth (I'm sparing you some technical details here, you'll understand) and, within a few hours, you'll have a beaker swimming with bacteria containing your desired gene, which you can then investigate to your heart's content. What we have here is, in fact, a lean, mean, custom-made DNA-generating machine. What's more, a slightly different manner of tinkering will have the *E. coli* manufacturing large amounts of a desired protein out of the gene that you put into it. In Chapter 6, we'll find out about entire industries that are based on this ability.

---

1. Please don't.

Fourth and finally, there's never an outcry about cruelty to microbes. Billions of these critters are killed daily, but I have yet to hear of anyone picketing for this to be stopped. Maybe it's the fact that *E. coli* is not technically an animal; or it's assumed that, because an *E. coli* cell has no nervous system, it can't feel pain and suffering (which may or may not be true, or relevant, to the ethical discussion); or it may just be that even the most ardent animal-rights activist flushes millions of them away per sitting. I suspect the most important reason is that *E. coli* don't complain or make distressing noises when mistreated, nor do they writhe in anguish (visibly, at least, because they aren't seen at all by the naked eye) and, in any case, they are neither cute nor furry.[2]

There's a lot more to be told about *E. coli*. The lab, however, is not the only place for meeting microbes.

Feel the Heat

Here's an experiment for you not to try at home.

Seal up your bedroom, fill it up to, say, neck height with boiling water, and sit inside it for a few hours.

Now, there are several good reasons not to attempt this: You'll die screaming in agony, I'll get the pants sued off me by your next of kin, and someone will have to clean up the mess afterwards.

But *had* you done it, you would have experienced what one particular microbe would consider a slightly chilly environment.

Strange creatures are often found in strange places. If you read a bit of the history of microbiology, you'll find plenty of cases in which

---

2. Ethical matters deserve a less flippant look, of course, but they are beyond the scope of this book.

it was said, "There can't possibly be any life in *that* place"—which is why it took so long for someone to go there and have a proper look. In this case, having a look would also have been quite a bother: This particular microbe lives in very inaccessible spots known as black smokers, which are tall natural chimneys (up to twenty-five to thirty feet high) that stick up from the ocean floor and spout super hot magma into the water from beneath the Earth's crust. We're talking up to 750° Fahrenheit here.

"Doesn't water boil at 212 degrees or so?" I hear you ask. Yes, in your kitchen it does; but at one and a quarter miles below sea level, the water pressure is so powerful that the hot water remains a liquid. Concentric circles of heat form around the chimney, with the hottest temperatures occurring near the chimney hole. Because cold ocean water surrounds the hole on all sides, cooler circular zones form around it: The zone nearest the smoker is insanely hot, the zone circling it is slightly cooler (very relatively speaking), and so on, until the outer zone, which is at ocean-water (i.e., near freezing) temperature. "The nine circles of hell" is not a very accurate description, but you get the picture.

If hell it be, we may reasonably expect all kinds of fire and brimstone in there—and we'd be right: The smoker itself is made of iron and sulfur minerals, and the water is thick with poisonous chemicals, from hydrogen sulfide to various heavy metals. To make things worse, it's very dark down there: Because no sunlight shines through at that depth, there is no opportunity for photosynthesis. Boiling hot *and* freezing cold; poisonous; armor-flattening pressure; pitch black—not exceedingly hospitable places, are they? Yet something calls them home, sweet home.

In 1997, a sample from a black-smoker area was found to contain a species of microbe that could withstand, indefinitely, exposure up to 235° Fahrenheit, and that was happily reproducing at 223° F (for a microbe, reproduction activity is a reliable sign that conditions are fine, as far as it is concerned). Temperatures below 194° F were, in fact, too cold for it. Named *Pyrolobus fumarii* (*P. fumarii*), which is Latin for "fire lobe of the chimney," it was the most heat-resistant organism ever discovered.

Heat resisters (hyperthermophiles) were discovered several decades ago and are very interesting to scientists, because such high levels of heat completely mess up "ordinary" organisms: proteins change form (boil an egg and see), DNA is a total loss, and the membranes of cells pop like bubbles; but these guys deal with it, no sweat.[3] Since their discovery, research has revealed some extraordinary engineering on these microbes that helps them function.

Hyperthermophiles are useful, too: An enzyme from the hyperthermophile *Thermus aquaticus* is essential for a lab technique called polymerase chain reaction (PCR), which is just about the most useful tool in biological research since the invention of the wheel. Briefly, PCR amplifies a DNA sequence, turning one copy of the sequence into millions. This enables us to study, analyze, and utilize that sequence. PCR is also the basis for every forensic DNA analysis that you hear about on the news (and, as a regrettable side effect, it has been the basis for the ongoing epidemic of *CSI* crime shows).

Back to *P. fumarii*. There's a bit more to this guy than just being able to take the heat. As we remember, the cycle of life depends on plants getting energy from the sun and nutrients from the soil, animals eating plants and other animals and then dying, microbes breaking down dead animals and plants into nutrients, and so on.

---

3. Microbes have all sorts of strategies for dealing with adverse conditions; for example, some can create protective shells (spores) around themselves and stick a sensor outside to wake them up when conditions get better, surviving in a dormant state until then.

The only noncyclical element involved in this process is the sun—it's the battery that powers the cycle. Every living creature is dependent on energy from the sun, and if it isn't there, the cycle stops; but *P. fumarii* (and a few other equally weird microbes) doesn't need the sun at all: It gets its energy from the heat and chemicals in its environment, via a series of complicated biochemical tricks whose details I will spare you.[4] It literally lives off the Earth and, in doing so, provides other creatures with the opportunity to live, by enabling the existence of an entire ecosystem: A varied collection of creatures that live near black smokers either have symbiotic relationships with the hyperthermophiles, or they feed on them. Other creatures prey upon those creatures, from wriggling tubeworms to poisonous-fanged snails that hunt iron-plated mollusks; so the vicinity of a black-smoker vent is full of life, while all around it, the ocean floor is a silent, barren wasteland.

You think 235° F is tough? In 2003, an even hardier microbe was found in the Juan de Fuca underwater mountain range in the Pacific Ocean, in a chimney complex about two hundred miles offshore from Puget Sound. Provisionally named Strain 121, after the highest temperature Celsius at which it can grow (about 250° F—and it can still survive at 266° F), it puts *P. fumarii* to shame. What's more, it uses iron instead of oxygen, and is therefore fully independent of the sun.

Various hyperthermophiles have been found in several different vent complexes all over the oceans; but how exactly did they get there? They can't just leave one vent and travel over to another, because they wouldn't survive the freezing journey. So how did they do it?

---

4. Actually, *P. fumarii* needs oxygen, which means it is a little reliant on the sun, because if the sun were to wink out one day, the oxygen supply in the oceans would eventually run out.

We don't know the answer, but here's something to think about: All hyperthermophiles, as well as other microbes that are resistant to extreme conditions (cold, acidity, and salinity, for example), belong to the group of organisms named archaea. They're not bacteria, and they're definitely not fungi—they're a different sort of creature. An archaean is as different from a bacterium as it is from us. They were called archaea because they're archaic, and although they live alongside us in the twenty-first century, they are thought to closely resemble the organisms that first developed on planet Earth; in other words, we're the ones who have changed, while they have stayed the same.

It may be that billions of years ago the conditions on Earth were pretty harsh—with sulfur and metal gushing into the ocean, and no oxygen—and the first life that developed was equipped to cope with that. Later, things calmed down, and life evolved to adjust to the changing conditions; but little islands of ferocity still remain and, within them, the unchanging relics of an ancient age still do what they have been doing for time out of mind.

A bit too soggy for comfort, perhaps? Let's head to Chile now to dry off a bit—more than a bit, actually.

A Sip of Life

The heart of the Atacama Desert in Chile, lying east of the Pacific and stretching into the Andes mountain range, is the driest place on Earth. It's dry, dry, dry. Rain is virtually unknown there. There are no rivers, no lakes, no water at all; and no water means no life. That's nature's way—no exceptions.

It is not surprising that this desert, although not especially hot (as deserts go, at least), is the most barren place on our planet. It is so devoid of life that NASA uses it to test its Mars landers because it is the environment on Earth most similar to the Martian wastes. No plants live here, no insects, not even any microbes. Nothing.

Yet in 2005, in an ancient, dried-up lake that's now no more than a collection of salty rocks, microbiologist Jacek Wierzchos found an extremely hardy type of bacteria living just below the rocks' outer surface.

What are these bugs living off? Where do they get the water they need? Why are they on the inside of these rocks, instead of at the top? *What's going on here?*

We don't know the whole story yet, but here's how we think it works: Even the driest places have some humidity, some moisture, in the air. When the temperature drops at night, some of it precipitates into tiny water droplets and stays on the ground—that's how we get morning dew. In the Atacama, there's usually very little moisture; however, some nights (not too often), usually just before dawn, there'll be a bit of dew hanging about. If it lands on a bit of salt, the salt sucks the moisture in (sprinkle a few drops of water on a bit of table salt, and see what happens), and water reaches our microscopic friends who, up to this point, have been dormant, lying patiently in wait for this occasion. Now they wake and quickly get to work, because this is the only window of activity they're going to get for a while.

First, they open up and sun themselves; like plants, they are photosynthetic bacteria (belonging to the *Chroococcidiopsis* family of cyanobacteria), so they absorb their energy directly from the sun. If they're lucky, and there's a bit more time, they'll do some self-maintenance chores; and on a very good day indeed, they might even get to go forth and multiply (which they do, of course, by dividing).

Fun in the sun is quickly over, because every moment means losing more precious water, so they close themselves up once again and lie among the glittering salt crystals on the bed of a lake that dried up millions of years ago.

Perhaps they dream of rain.

Slime City

I misled you earlier, when I was talking about *E. coli.* The *E. coli* we work with in the lab is not exactly the type we find in our guts. Over time, the lab variety has been bred to become more user-friendly—in particular, to lose its tendency to secrete large quantities of a vile, sticky substance that interferes with its handling. This is the story of the wondrous things (which we call *biofilms*) that are made of this vile, sticky substance. They're something that you yourself spend a lot of time trying to get rid of—and a lot of money, because you aren't able to.

We tend to think of microbes as swimming around in watery places, but that's not always the case. Frequently, a microbe will come across a solid surface and latch on to it, with the help of some sticky proteins on its surface. The microbe will then start to ooze, secreting a sugary substance that forms a slimy shell around itself, which other microbes stick to. They start secreting, too, and pretty soon there's a whole heap of secreting microbes—a busy, thriving jumble of them, layer upon layer, surrounded by a protective outer matrix, each occupying a teeny cavern that has formed within the ooze.[5] Have you ever been to an overpopulated city like Hong Kong, a Middle Eastern casbah market, or the less salubrious parts of New York City? It's a little like that, only with more ooze.

---

5. The word *matrix* has acquired a lot of symbolic significance. Let it go this time—sometimes a matrix is just a matrix.

It's now estimated that most microbes on Earth live in biofilms; but we still know surprisingly little about them, because that kind of dense, shut-in structure is very hard to study. Scientists prefer to deal with easily measurable things, like bacterial cells that float around individually in liquid. When cells start clumping together, sticking to things, and refusing to budge, it's very confusing: you can't determine their concentration properly, because they're not evenly dispersed; meanwhile, it's hard to get at them, because they're protected by the outer matrix. You can't even know what exactly it is you're studying, because there are different kinds of microbes mixed up in there, all with different properties. This, as we will see later, also makes it hard to get rid of them. We're not talking about any special, ultraresistant bugs here, either—but once they arrange themselves in this way, they're bastards to deal with.

And arrange themselves they certainly do: Biofilms aren't just a bunch of bugs, one on top of the other. Usually, many very different species live together side by side (very culturally diverse, the biofilm community), communicating with each other. They do this by exchanging all sorts of chemical signals. Totally different sorts of creatures understanding each other's languages—isn't that something? Biologically, a fungus "talking" to a bacterium is like you exchanging pleasantries with a shrub. It's hard for us to tell exactly what they're "saying," and anyone who has observed any part of nature at any length could guess that conflict is definitely part of this game; but the end result is a thriving multicultural habitat, so they're doing something right in there.

Biofilm structures have plumbing, too: Channels run through them that supply the resident microbes with necessary water and nutrients, as well as good communication routes (because signaling molecules also circulate throughout them). All in all, it's not as ramshackle an affair as it may look from the outside: Microbes in biofilms don't grow every which way—they arrange themselves so they don't block up the water channels, and they differentiate themselves so that the outer layer is tougher and more resistant than the thriving inside layer. In short, they're organized.

But biofilm inhabitants don't always stay put: Adventuresome microbes often detach from the main structure—either alone or in little clumps—to seek their fortunes further afield (or further downstream, in their case, because they go with the flow). They'd probably whistle "Westward, ho!" as they went, if they had anything to whistle with.

Biofilms can be found anyplace where there's moisture and a surface to cling to: they make rocks in rivers slippery; they clog up drains; and they're also a terrible problem in hospitals, where they cling to catheters and refuse to budge. Because they are constantly sending out new colonists, they then create a chronic contamination problem for people who need tubes stuck in them for any reason.

They also form inside our bodies: One particular sort likes to grow on the valves of our hearts and causes serious inflammation problems, and it's one reason why biofilm research is gaining a lot of momentum. In another form (plaque) they stick to teeth, feed on any sugar they can find there, and secrete acid that eats away at the tooth.

To be fair, scientists and environmental engineers are also putting biofilms to good use, cleaning up wastewater and other nasty stuff. With their diverse species and abilities, they can carry out a concerted effort that a single species just can't manage.

Humans may view biofilms either as useful or as a source of trouble. But what's the point of the biofilm to the microbe itself? Why should a microbe go to all that trouble to hang on to a surface or to form such elaborate structures upon it?

One answer is (to quote the late, lamented Sir Edmund Hillary, conqueror of Mount Everest), "Because it's there." As we've seen throughout this chapter, if there's somewhere to live, something will find a way to live there, and solid surfaces in wet environments are no exception.

It's also much easier for microbes to withstand the force of flowing water by sticking together, and the external matrix doesn't only keep things stuck together; it keeps things out, too, protecting its inhabitants from all sorts of stuff, including toxic metals, dehydration, and ultraviolet radiation. This may be one reason that microbes build biofilms.

Biofilms are also more resistant to antibiotics. It's not clear exactly why, but it might, again, be due to the protective qualities of the matrix. Another theory is that a biofilm may contain small subpopulations of microbe cells that can resist antibiotic attack, either because they are naturally resistant, or because they are dormant (an inactive microbe is much less vulnerable to most antibiotics). When the biofilm gets whacked by antibiotics and most of the cells inside succumb, these small subpopulations may start multiplying rapidly, restoring the previous state within a relatively short time. Good for them, a headache for us.

Biofilms are here to stay, so brush your teeth after meals, and visit your dentist regularly—for now, that's the best we can do to keep them under some sort of control. When all is said and done, you just can't beat city life.

The X-Bug

The X-Men (of comic and film fame) are mutants, which means they have an alteration in their genetic code that gives them superhuman powers, such as death-ray eyes and weather control. For some reason, they also seem to fly around a lot, and they have killer bodies.

Way cool. Sadly, as much as we would like them to be, mutations that allow us to defy physical laws such as gravity are not possible in reality.[6] But mutations do occur, constantly. In real life, most mutations are bound to worsen the microbe's situation, which is why the superpowers of this next bug are directed at avoiding them as far as possible.

Deinococcus radiodurans (*D. radiodurans*—but we'll call it D-Rad here, for added funkiness) is probably the closest thing we have to a real superhero; not for nothing was it nicknamed Superbug and Conan the Bacterium. Its superabilities include unbelievable resistance to radioactivity; immunity from heat, cold, and acid; and the ability to survive in outer space. D-Rad was discovered by accident, in an agricultural lab in the 1950s, when a can of meat was blasted with radiation that was supposed to kill anything living inside. When scientists investigated later, they saw that the meat had spoiled, and growing happily on it was D-Rad. Had it been a movie, D-Rad would probably have been wearing full-body tights and bellowing, *"Mwahaha!"*

To give you some perspective, humans would die from exposure to 400–1,000 rads of ionizing radiation. Most mammals are about as resistant. The humble cockroach, hailed in popular myth as radiation-proof, can tolerate ten times as much, as could the *E. coli* in your gut. One type of particularly hardy parasitoid wasp can take up to 185 times as much. But D-Rad puts us all to shame: It thrives happily after being zapped with 1.5 million rads, and can apparently withstand twice that. Mwahaha, indeed.

---

6. It's a shame. I'd like to be able to stretch time so I could sleep more. I'd be known as Snoozerman, and I'd wear purple pajamas and a cape. Fluffy slippers, too.

D-Rad's capabilities raise two interesting questions, both still under intense investigation: How is this resistance achieved and for what purpose?

The answer to the first question, as far as we understand it today, goes like this: Damage and death from ionizing radiation occurs mostly because the radiation damages the DNA inside the cells. The worst type of damage occurs when the DNA strand breaks completely.

Every living thing has some DNA-repair capabilities (your body is constantly being repaired by several systems—without them, you'd get cancer and die within weeks). D-Rad has those same capabilities, and many more.

First, its DNA is held tightly together in a special ringlike structure called a toroid, so that if its DNA does break, the fragments stay held in their original places and don't drift away (sort of like the shatterproof glass that car windows are made of). As a result, it's a lot easier for the puzzle pieces to be put back in the right order.

Second, once this shatterproof DNA is done with its *Terminator 2* stunts, a self-repairing system sets in. Each bug has between four and ten copies of its DNA, kept in four separate compartments, and only one or two copies are active at any time (when DNA is in use, it is more "open" and susceptible to damage). After the DNA within each copy has been repairing itself for a while, copies start going from one compartment to another, and the cell's repair enzymes start comparing every two copies, using each as a template to fix the other. In this way, the copies with greater damage are repaired, and the bug can go on living normally.

Now to the second question—what use is such a complicated system to this microbe? Evolutionary pressure dictates that if this system wasn't helping D-Rad survive, it would have been stripped down—or it would not have developed at all. So how come this microbe has built up such a complex system for dealing with radiation, when until very recently, there was no place on the face of the Earth with anything near those kinds of conditions?

"Why" questions are a bit difficult to answer when dealing with microbes (who are you going to ask?); but before we turn to science-fiction ideas (underground caverns filled with radium? ancient submerged nuclear reactors from a long-gone civilization?), one possible explanation is that this resistance might be a defense against something a lot more common—lack of water. Extreme desiccation causes just about the same kind of stress and damage to DNA that radiation does. Long ago, this species may have developed a mechanism for dealing with very dry surroundings that, incidentally, makes it immune to nuclear radiation.

All these capabilities are very convenient for us humans, because D-Rad is a superhero that uses its powers for good rather than evil. It is harmless to humans and animals, it lives peacefully in the soil, and it may well help us clean up the terrible mess that we have foolishly created for ourselves through our development of mixed-waste sites.

Microbes, you see, eat anything. They even consume man-made chemicals and plastics, which is very good news for us, because we don't know how else to get rid of all the waste and contamination that we've created. If we can find, or develop, microbes that like our chemical waste so much that they hungrily devour it, we will solve a very big problem for ourselves and for our environment. This field of research—bioremediation—has been gaining momentum recently and is explained further in Chapter 6.

In past decades, when the most important thing in the world for some governments was manufacturing lots of nuclear weapons very quickly, countries accumulated huge amounts of mixed waste—waste that is both toxic and radioactive. These astoundingly dangerous chemicals were stored in containers that started leaking after a decade or two, contaminating the ground and water around them. Ugh, humans.

There are now more than one hundred such sites in the United States alone. We might have been able to decontaminate the sites using microbes that break down toxic chemicals, but they would all die from the radiation before they managed to do anything.

Never fear—D-Rad to the rescue! Scientists are now trying to take the genes from the bioremediating microbes (which are poison-eating superbugs in their own right) and insert them into D-Rad. If this works properly, it would be an elegant solution to a very knotty problem.

A final thought about this bug: As I mentioned, it is also very resistant to vacuum conditions, so it can survive a trip through outer space, say, in a meteor or spaceship. This means that we have to be very careful about items we send to other planets—we risk contaminating them with our earthly bugs. Some researchers go further and suggest that it might actually be the other way around: that long, long ago, from some planet far, far away, an ancient bug resembling D-Rad was riding on a meteor and landed on a lifeless planet. It started multiplying and evolving, and that's how life began here on Earth.

That's probably not the way it happened, but a bit of a mysterious past has never hurt a superhero.

BONUS TRACK #2

Undead Gummy Bears

We're now proud to present our bear act—not just any bears, ladies and gentlemen: These are water bears. Give 'em a hand!

Formally known as tardigrades, these creatures are not, strictly speaking, microbes—they're actually animals. They're multicelled and have proper organs, and they're distant relatives of ticks and mites; but I have allowed myself to include them here because

1. They are very, very small.
2. They are awesome.

This large family of tiny creatures has been around for hundreds of millions of years, and has been known to humans since 1773; however, because they're quite harmless, most people have never heard of them (trouble equals recognition, after all—just ask Norman Khan, Genghis's unsung brother).

They were nicknamed water bears because they resemble bears in their appearance and in the way they waddle (though they walk on *four pairs* of feet). Like bears, they have claws, eyes, muscles, and nerves, and they move by clawing onto things like tiny mountaineers; though with their psychedelic colors and sticky-looking skin surface, they sometimes look more like gummy bears.

More than a thousand species of water bears have been recognized, with probably thousands more waiting to be discovered. They can be between 0.1 to 1.2 millimeters in size, depending on the species, which means that you can just about manage to see the biggest ones with your naked eye, if you're lucky—the problem is they're nearly transparent, and hard to spot.

Unlike bears, they breathe through their skin, which has to be constantly wet (that's why they're called water bears). They live just about anywhere that water is found, including very high mountains, oceans, Antarctic regions, and tropical rainforests, but can usually be found on mosses and lichens. Wherever you happen to be reading this, there is probably a water bear not too far away from you.

As for food, water bear species have different tastes: Some can eat bacteria, and others can suck the juice out of plant and animal cells with teensy-weensy fangs. In turn, they get eaten by other creatures, including amoebas and other species of water bears.[7]

---

7. This is the first time I'd heard of an animal being eaten by a single-celled creature. Frightening thought, isn't it?

Water bears also resemble bears in their ability to hibernate; but in this regard, the tiny tots far surpass actual bears. In winter, a regular bear may go into a long hibernation, during which its metabolism slows down so it won't use too much energy. When conditions get tough and inhospitable for tardigrades, especially when their environment dries up too much for their taste, they can enter a state of cryptobiosis (derived from the word *crypt*), which means that they can bring themselves to a mode of having next to no metabolism—a state formerly known as death—and stay like that for long periods until revived.[8]

The water bear accomplishes this extraordinary feat by drying itself up almost completely, replacing nearly all of the water in its cells with a sugar called trehalose. It can stay like that almost indefinitely, until some water drops on it, whereupon the water bear regains consciousness within a few minutes or hours, and goes about its business as if nothing had happened.

What's most astonishing is the water bears' resilience while in this suspended "undead" mode: Various tardigrade species can survive in temperatures as high as 304° F (infernally hot) and as low as -459.67° F (as cold as cold can be).[9] They can handle high pressure, toxic chemicals, and even vacuum conditions; and then you sprinkle water on them, and wait a bit, and they get up and waddle away, cute as cupcakes.

Researchers are beginning to get interested in these critters, because the trehalose trick, if we find out how to do it, may prove very useful for keeping living things ready for use but dormant for long periods—things like organs that are waiting for transplant, or astronauts who are going on trips to faraway planets. Think about it: *powdered astronaut—just add water!*

Most of the research on tardigrades has been concentrated in Europe. Not much is known about North American tardigrades, and because they're all over the place, you could do a little original research at a moss near you, and you'd have a fairly decent chance of finding a new species of animal, completely unknown to science.

Bear that in mind.

---

8. In a well-known case, a water bear was revived from moss after one hundred years of cryptobiosis.

9. If you're wondering whether this contradicts Strain 121's claim to the world record for heat resistance, don't forget that these water bears, unlike the hyperthermophile microbe, are not living normally in these conditions—they're suspended, inactive, waiting for better times; so they're surviving, not living.

CHAPTER 3

Bugs on the Move

Have you ever visited a ghost town? Looking around, you see nothing but wasteland, yet people once worked and raised children there and called it home—and then everybody left. Why did they leave? Why did they come there in the first place?

The answers always have to do with some sort of resource. Boom towns often develop after the discovery of gold, for example, while other precious or useful resources (metals, oil, and timber, usually) also tend to get people clambering from all around, keen to get a share before it runs out. The most basic resources are good farming land and water, but even a good position—on top of a defensible hill, say, or in a sheltered bay—can be a resource in itself. Obviously, individual people have complicated reasons for going to one place rather than another, but if we look at the big picture—at mass movements of people over long lengths of time—the patterns are there.

Microbes? They're after exactly the same things we are: water, usable energy, and materials. It's therefore not surprising that the dynamics of their habitation are similar: Locate something you can use, use it, and move on when it's depleted—or when you find, or hope to find, something better.

This last point is interesting: If everything's all right as it is, why bother moving at all? What's the underlying evolutionary rationale for a microbial species to develop what we'll see are very complicated and costly mobility mechanisms? The reasons are clear enough, but not immediately intuitive. I'll be cheeky and sum them up using a most unscientific word: *hope.*

Hope Springs

In humans, hope is a very tricky thing to analyze, but let's try to approach it using the example of investments.

Suppose I have some spare money (truly a hypothetical example, to be sure) and I wish to keep it and, if possible, make some more. What I need is what smart people with even smarter suits call an investment strategy. Investment strategies can be high risk or low risk; the higher the risk, the bigger the payout should be. Stocks are usually higher risk than bonds or term deposits, and putting everything you have on one single stock is even riskier. An extreme example of a conservative investment strategy would be to buy a lump of gold or iron, and to sit on top of it with a loaded shotgun until you die (the trust-no-one school of finance)—this, I am told, isn't expected to give any significant yield on the investment.

A surprisingly popular example of the other extreme is lottery tickets (the may-fortune-smile-upon-me paradigm), wherein a small investment may, ultimately, yield huge profits but, not to put too fine a point on it, doesn't.[1]

---

1. Gambling is sometimes called a tax on stupidity. I agree unreservedly, but I still buy my annual lottery ticket as a kind of dreaming license. A certain measure of stupidity is an integral part of being human, after all.

Investments can be diversified, or they can be focused on one type of asset. For the long-term client, any investment advisor (without hastily printed deeds and a getaway car, anyway) would suggest at least some diversification, because it makes good sense. It's hardly surprising, then, that living organisms—microbes included—employ the same modes of behavior: Math is math, and even though microbes don't consciously calculate odds, the same rules of statistics and probability apply everywhere. As such, a microbe faces an ageless dilemma: Should it stay or should it go? Should it stick it out, or try its luck elsewhere?

As I won't tire of repeating, a microbe doesn't think; the decision is taken on an evolutionary level, and the individual microbial cell doesn't have a say in the matter. But because a species of microbe that cannot move at all will, most likely, face serious resource-depletion problems at one stage or another, and because, in some cases, golden opportunities *are* lurking just beyond the horizon (perchance in the next puddle, or on the next poodle), a microbial species will usually hedge its bets by evolving an overall strategy that includes at least some element of moving about.

Many types of microbes don't move actively at all, relying on passive forces—from air and water currents to more elaborate tactics—to carry them wherever they need to go. Others don't move around much themselves, but use various ingenious means to get their next generation dispersed. Here's the story of one such microbe, which I particularly like:

Note: This next section contains an offensive word. If you come from a very good family, kindly refrain from reading lines 1, 2, and 50.

Cow-Crap Cannons

Life for *Pilobolus crystallinus* (*P. crystallinus*) is shit. Which is exactly how *P. crystallinus* likes it—cow shit, preferably, but any grass eater's dung will do.

You and I may think that residing in dung is a disgusting way to spend your days, but crap is a very wholesome home, if you're not too picky: There are plenty of nutrients left there, even after the cow is done digesting, and it's nice and warm, so there's always plenty of action around the area, with insects, worms, and heaps of germs joining in the fun.

One cowpat is nice, but even good things don't last forever, and *P. crystallinus*—aw, shucks, let's just call it PC—wants its offspring to move away to newer and greener (well, browner) pastures. You want the kids to have their own shiny, succulent, new feeding ground, otherwise everybody will be bunched up together, and food will run out very quickly.

But how to get them there? Moving from dung to dung is all right if you're a fly, but if you happen to be a fungus, like PC, you've got a bit of a transportation problem. A common microbial technique is to hitch a lift on an insect. You produce some spores and cover them with a sticky substance, the fly comes to feed, the sticky sack gets stuck to its body, and away it flies to the next dung heap, taking the sack with it.

Simple enough, but PC doesn't trust flies, and besides, it wants its future generation to have the best, freshest dung there is, which means going to the source—yes, PC wants to get its spores *inside* a cow.

Now cows may not be the cleverest of animals, but they're plenty smart enough to know not to eat what they've just pooped. That makes good sense (except to some dogs, but never mind them); so PC needs to get its spores as far away as possible, onto the fresh green grass. How? It turns itself into a water cannon.

While its spores are maturing and getting ready to be scattered, PC grows a stalk with a sac at the top (called a sporangium). PC pumps water into the sporangium so that it accumulates inside the sac, creating high water pressure. When the spores are ready to go, the water pressure is released, and the ensuing blast sends them flying. This artillery barrage can reach a maximum height of six-and-a-half feet, and can land about about eight feet away—that's not bad for a half-inch-long cannon.[2]

The spores don't land just any old where, either. PC is a smart, hi-tech fungus, and it aims them toward the sun, using a light-sensitive area found just below the tip of its stalk; that way, there's a better chance for the spores to land in a sunny patch, where the grass is growing well—exactly where a sensible cow might want to come and have a spot of lunch.

A cow will now come along and munch away at a patch of that grass, and in with the mouthful will go a spore. Its outer casing is hard, so it doesn't get chewed or digested; it just goes patiently through the cow's digestive system (and a cow has a *lot* of digestive system), until . . . *splat* . . . out comes a spore, completely encased in a fresh, untouched, appetizing, organic, delicious, nutritious, sweet-smellin', just-like-home piece of shit.

Another movement challenge that microbes face is the matter of how to establish themselves inside a host organism, rather than just going in one end of the digestive tract and out the other.[3] Then, of course, there's the problem of how to move about inside without being hassled by the fuzz. But we're moving into darker territory

2. I asked an engineer who maintains hydraulic systems on planes how strong the pressure needed to be to do this. He just said, "Very."

3. Physiologically speaking, the entire tube from mouth to anus doesn't count as being "inside" the body at all. A human being can thus be regarded as a very complicated hollow tube.

now—much as we may admire this next microbe's smooth moves, he is 100-percent-pure bad guy.

Terrorist Cell

Listeria monocytogenes (*Listeria*) is one nasty bug. Many microbe species develop remarkably ingenious ways to evade the body's security forces, and it's probably one of the cheekier ones that we know about.

Humans can catch *Listeria* from eating contaminated raw food (it's quite common in the environment, and in wildlife and livestock), whereupon it can cause plenty of problems, including diarrhea, spontaneous abortion, and death. It's a major headache for medical and food-safety authorities, and no less of a headache for our immune systems.[4]

Once *Listeria* gets inside a digestive tract, it goes to work. In phase A, it smarms its way into an intestinal epithelial cell by using a special molecule securely fastened to its surface that is able to communicate with a molecule of the host cell, and tricks it into starting a chain of molecular reactions.[5] The reactions cause the host cell's own membrane to form a bubble around the germ, which can then gain entry into the cell without being detected.

In phase B, once inside the cell, *Listeria* sheds its enveloping bubble disguise and starts multiplying. Once its numbers have grown so that the host cell has become, in its opinion, too crowded, it goes into phase C, which involves penetrating other cells in the body, while

4. That said, a healthy immune system can usually cope with a *Listeria* infection—the problems begin when our immune defenses are not at their strongest.

5. The manipulated protein molecule, E-cadherin, normally has a perfectly innocent role in mammalian epithelial cells.

keeping out of sight of the immune-system components that constantly patrol the body's highways and byways in search of just such infectors.

As it turns out, about two weeks after learning about this microbe, I read something in the news that offered a very similar solution to a very similar problem. The problem in this case was a military operational one: How can an infantry force move, however slowly, through a very hostile, very crowded urban area without exposing itself to enemy fire? The military solution was simple, brutal, and effective: Move through the walls. After gaining control of a building, the soldiers would blast open a hole in one of its side walls and move into the adjoining house without needing to come out onto the street. Leaving aside for a minute the huge, horrible implications such a tactic has on the people living in these buildings, one must admit that it does achieve its operational purpose.

Listeria employs the same method: Instead of risking exposure to hostile immune forces, the germ once again hijacks its host cell's systems, this time taking control of the host's actin-polymerization machinery. Actin is a common protein in our cells, from which long rods—actin polymers—are created. These rods are rapidly built up and taken apart as needed, and they serve to keep the cell's shape intact and to move material around inside the cell—picture an elongating fireman's ladder with a fireman at the end. *Listeria,* however, takes over the controls and causes these actin polymers to build up behind it at a rapid pace, creating structures known as actin comet tails, which propel the germ with some force into the host cell's membrane and into an adjoining cell's membrane, whereupon the whole process starts over.

Many and varied are the microbes that infect us and the ways that they manage to stay alive. *Listeria*'s technique is just one example;

but host manipulation by microbes is by no means confined to cellular machinery. A microbe infecting a host will, sooner or later, run out of cells to infect. An efficient microbe must develop strategies for jumping from one individual host to another (and another). We've already seen how a fungus uses itself as a cannon, but why not use someone else for the hard work?

Travel Arrangements

It's cardboard-box-hunting time for me again—we're moving house. Our present abode is very nice indeed, but it won't be able to accommodate the oncoming population expansion that's due to occur for our family shortly; so we pack, and Scotch tape, and mark clearly, and wrap, and heave, and perform all the usual tasks that accompany the transfer of lives and belongings between places. Because I am an indifferent housecleaner at the best of times, an important part of the ritual is always the ultrathorough cleaning and dusting that becomes necessary when furniture is moved from its customary place for the first time in many years. It is not surprising, then, that dust and moving house have become linked in my mind and that, as I drag another crate of books out the door, my nose tingles, and I stop to salute one of nature's great movers: the sneeze.

What a piece of work is a sneeze: so useful to the body in cleaning out its air pipes and preventing dust particles from accumulating; so commonplace in occurrence, and yet so gratifying to perform. It is the only bodily function that unfailingly earns us blessings and wishes of good health from our fellow men; and though it be unspectacular to watch, is it not a wonder of muscular coordination and power to send sprays of bodily fluid a distance of ten feet away, at a speed of up to a hundred miles per hour? Most assuredly so, and this last fact has not gone unnoticed by those of a microbial bent squatting among the alveoli of our lungs, or hiding in the folds of our windpipes. "Ten feet, you say?" I imagine them asking. "Why, that is to us as a free ride to Fiji is to you! Pack your suitcases, everyone!"

Sneeze Airlines has always been a popular mode of microbial transportation: Long before the appearance of mankind, microbes were already whizzing between respiratory systems, being forcefully expelled out of one animal and inadvertently inhaled by another.[6]

Sneezes are triggered by the sneeze reflex. Irritation of the nose or airways activates the trigeminal nerve, which then signals the sneeze center of the brain. This center then coordinates all the muscle responses involved in sneezing, and hopefully the cause of the irritation is ejected from the body.

What a splendid way to move about—a tickle, and you're off! The entire machinery is already set up and waiting for the cold or flu virus to do nothing more than be noticed by the immune system: Inflammation gets underway, the airways become clogged with the usual inflammatory debris, irritation is produced, and voilá—a sneeze takes our virus elsewhere, to conquer new, uncharted noses.

Could this process be considered an act of manipulation by the virus? The virus benefits from something that is already happening anyway (the inflammatory process), but does it mean to? Does it contribute anything toward the process? Pathogens (disease-causing microorganisms) frequently employ mechanisms to avoid immune reactions; is this a case of a virus deliberately getting into a scrap?

The question is totally meaningless, on one level: Viruses don't *mean* to do anything; they're nothing but a bit of DNA inside a

---

6. Totally useless science fact: Cows sneeze seven times as much as humans.

protein envelope. On another level, asking whether viruses have evolved a way to actively exacerbate the inflammatory process is a slightly more meaningful, but very difficult, question to answer. Because inflammation is the result of the body fighting against an invader, how can we tell whether a virus is actively inducing inflammation (to trigger sneezing), or simply fighting for its own survival as hard as it can?

I don't know the answer. I'll stick out my neck here and suggest that no one does. I'll stick it out even further to muse that perhaps there is no answer—just two complementary ways of looking at the situation. On an evolutionary level, a viral species will strive toward an optimal balance between not making trouble (for survival and reproduction inside the host) and making trouble (for spreading to another host). That balance is not fixed: It changes continually, responding to the virus's environment (our airways and immune processes), which is also in a continual state of change and response; and thus we continue, horns locked eternally.

Too philosophical? Let us balance this with a gross digression: The dried-up remains of respiratory-inflammation processes are found in the nose. Commonly known as boogers, they include cellular debris and bacterial and viral remnants.[7] There's a respected medical opinion that suggests that eating your own nose pickings may actually boost the immune system and be good for your health. Other doctors, I should add, disagree. You may keep your own counsel as to what you choose to do with this information; but, for your own safety, avoid telling this to anyone under twelve when within earshot of his or her parents.

7. The smoking hulks left over from the battles of the nostril, enveloped by mucus.

Traveling south from the nose, another very popular exit route out of the human body for microbes is via the back door. When the microbe causes diarrhea, that exit is usually very speedy. Diarrhea is a condition that causes us much suffering, and results in a terrifyingly large number of deaths worldwide (especially in children, who are more susceptible to the rapid dehydration that it causes). For a variety of microbes, it is a very useful aid to dispersal, because their chances of reaching and contaminating a drinking-water source are good. Once again, a healthy bodily cleansing process (the runs are very useful in flushing out bad food, for instance) is subverted by microbes for their own odious purposes.

Inventing the Wheel

We've been talking nonstop about infections for a while now, and it's beginning to sound as though infecting other organisms is the only thing that microbes ever do. The great majority of microbes, however, don't bother infecting anything at all and, instead, travel around using their own movement devices.

A large variety of microbes develop a long, slender projection called a flagellum (sometimes several times longer than the rest of the microbe), which constitutes their method of propulsion. It is made of more than twenty separate components that are assembled outside the microbial cell (with the help of another thirty or so other components), and it contains, as an integral part of its structure, what is probably the first functional wheel ever, predating the Bronze Age by a few hundred million years.

Flagella appear in many forms and sizes, but the basic shape of the bacterial flagellum is very common, so I'll use the much-studied *E. coli* as our model.

Most visible is the flagellum's long, whiplike structure, which is composed of a protein called flagellin. At the tip of it is a capping protein, while at the other end are hook proteins that keep it bolted to the cell. Its base structure has a whole bunch of proteins acting as bearings, channels, and all kinds of things (technically speaking) whose job it is to keep the flagellum moving, and to build it up and repair it while it works. The whole thing is basically a motor that runs on protons and goes around in a circle, rotating the whip like a propeller, in order to move *E. coli* along.

We can see a lot of variations on this theme in other organisms. Archaea, for example, have something that does the same thing, looks just about the same, but is actually very different. Eukaryotes (a diverse group of organisms, encompassing amoebas, oaks, frogs, and you and I) also have flagella that are put together very differently, but perform those same tasks—the most famous example of these is sperm, which rotate like crazy to get ahead in the race to fertilization and life.

E. coli flagella have an interesting switch arrangement that allows them to reverse or rotate the other way. The switch is connected to equipment (more proteins) that senses chemicals and conditions in the water that the cell is swimming in, and works the propeller accordingly. This effect is called chemotaxis: Essentially, an organism's ability to move toward a source of molecules that it is attracted to, such as food, or away from a source, if needed. It's the microbial equivalent of a nose—it sniffs out good smells, and directs us away from bad ones—and it may give us a clue about how our sense of smell originated. There's also phototaxis (a microbe's movement towards or away from light) and, lately, other kinds of bacterial senses have also been discovered: magnetotaxis, a microbe's movement in response to magnetic fields; and elasticotaxis, a microbe's movement in response to physical tension.

The first thing to note about flagella is that there's no brain governing their reactions: just a few proteins sensing and a motor moving.[8] Mindless attraction.

The next thing to note about them is what they do when they're *not* moving toward something. Back to our *E. coli,* for example: It will chug over, suddenly switch to reverse for a second, and then chug away in the opposite direction. It's just like what happens if you pull repeatedly on the hand brake while driving at speed on a wet surface (don't forget, we're in the water here): The bug moves forward, spins off in a new direction, and then the whole thing happens again. This is known as tumbling behavior, and scientists figure it's a way for the bug to check out its surroundings, cruise around, and then head off in a new direction in the hope of getting a whiff of something good. Mathematicians call this "random walking," and they've proven that a microbe can, and in due course will, eventually get anywhere this way.

Single-celled creatures use flagella to move themselves around, same as the other microbes; but in more complicated organisms, flagella-like structures called cilia are often used to move *other* things around. In our airways, for example, we have masses of cells lining the windpipe that move their cilia in coordinated waves to shift mucus and nasty foreign stuff outside, where it belongs, keeping our lungs nice and clean. Inside our ears, cilia are receptors for sound, moving when soundwaves hit them, and transmitting the signals onward. We can find cilia being useful in all sorts of locations, including

8. In humans, there are only a small number of neural shortcuts that don't involve the brain. An example: When you touch something hot, your hand moves before the message *"Owwwww!"* gets to your brain, which then processes the information and sends back a response. If you'd waited for your brain's message, your hand would've been nicely grilled. The sensory neurons in the hand communicate directly with the motor neurons, which trigger a quick muscle spasm. Sort of an idiotproof feature.

our eyes, nose, and kidneys—strange places for former propellers to be, but that's nature for you.

Irreducibly Boring

Before we move on, allow me a short political digression. The bacterial flagellum is truly a marvel of natural engineering, so much so that we're still not sure exactly how it came together. This last fact has, in the past few years, provided some unexpected publicity for what is otherwise a rather obscure field of inquiry. The complexity of the flagellum's structure and function is currently providing argumentative ammunition for supporters of intelligent-design theories, who claim, as far as I can gather, that because we don't yet understand how the flagellum evolved, we never will, so we should give up trying and conclude that a higher force has been at work here.

I won't go into this debate in detail here, because we're trying to have some fun, and politics is a concept I find to be diametrically opposed to the idea of fun—but I do need to explain why this is politics and not science. This is a battle of doctrine. Intelligent-design supporters argue that "we'll never know" is a legitimate scientific claim. The rest of the scientific community says that such a statement is precisely what science *isn't*. The debate itself is not science; it's a power struggle over the meaning of the term *science*, with a nonscientific view attempting to cram itself forcibly under the heading of science in order to gain respectability. This is brand-name dynamics, not different in principle from an herbal diet pill claiming to be "scientifically proven" in order to boost sales. It also, ironically, shows how much public weight the science stamp of approval now carries, which I'm not sure is always a good thing.[9] The debate is not concerned with the

9. That was me not deliberating on this subject. Oh, dear.

flagellum itself (which everyone concerned seems to approve of), and besides, I find the entire thing rather silly and tedious, but I would like to relate something from an unexpected source that may bear on the subject.

A renowned rabbi once explained why he (and orthodox Judaism in general) does not attempt to prove the existence of God by logical debate, as other religious leaders sometimes have. What happens, he asked, if an error is later found in the proof? Would that negate the existence of God, or lessen the force of my devotion? Belief in the almighty is an axiom, not a logical conclusion.

Although I don't share the rabbi's religious views, I do agree with him on this particular point. A few decades ago, we could not have adequately explained how anything came to be. Now, we cannot fully explain the evolution of the flagellum. With the rate of advance in scientific insight that we're currently seeing, anchoring one's worldview to the irreducible complexity of the flagellum seems like a short-term bet to me.

The Amazing Xanthus Brothers

Part 1: The Streetcar

Little rotors are by no means the only mechanism that microbes use to get around. *Myxococcus xanthus* (*M. xanthus*) is a case in point. It doesn't normally get a lot of publicity, because it has little to do with humans, but that's our loss; this is one heck of a germ.

M. xanthus lives in the soil, so it moves over solid surfaces, rather than swimming in liquid, and it uses a remarkable double-engine propulsion system to do so. Each *M. xanthus* cell has two motors—a puller and a pusher—which it uses to move over the surface (or over other *M. xanthus* cells, which we'll get to in a moment).

The first motor, called the S engine, is basically made of pili. Like flagella, these are whiplike structures, but they're made of different proteins and have very different functions (including, incidentally, sex—but you don't care about that. Oh, you do? All right, we'll come back to it). In *M. xanthus,* these pili can be up to six times as long as the bacterium itself, so that they stick out one end of it. The tip of each pilus sticks to the surface, and the cell then heave-hoes itself forward, like a horizontal mountaineer pulling on ropes with grappling hooks.

At the other end of the cell, the nozzles of the second engine, the A engine, secrete a slimy substance that apparently helps push the cell away. Scientists were puzzled by this mechanism, and several explanations were proposed, including that it was something I can only describe as a slime jet: A slime-squirting propulsion that moves the bacteria in the opposite direction, in the manner of a jet or rocket engine. Recent findings replaced this literally repulsive idea with a view of the A engine as a neat little arrangement of adhesion points that are constantly being formed along the body of the *M. xanthus* cell, which allow it to push against the outside surface and move along in a corkscrew motion. The adhesion complexes then disassemble at the back end of the cell. This is rather elegant, but it doesn't really explain the need for the slime. Hmm.

Occasionally, *M. xanthus* feels like going the other way, so it shifts into reverse gear: Most of the machinery for the motors is found at either end (not unlike a streetcar), so the S engine shuts down at one end—the pili retracting into the body of the cell—and starts up at the other end, where new pili sprout. The A engine does likewise, and *M. xanthus* glides away serenely in the opposite direction, leaving a trail of slime behind it.

Part 2: The Swarm

M. xanthus are predatory bacteria that hunt other microbes in swarms: Millions of *M. xanthus* cells form a single hunting pack and use their quorum-sensing abilities to find out where most of their "brothers" are headed and follow them. The slimy trails they leave behind allow other *M. xanthus* to move faster along that same path. The result is a sci-fi-esque, alien-looking blob that grows in every direction at once until it spots its prey, extends an "arm" toward the helpless colony of bacteria, and engulfs it. If it were large enough, we'd probably hear a *gloop* noise.

How do blobs sense things, anyway? Bacteria have no eyes or ears, so *M. xanthus* uses its other senses to find its prey: either by chemotaxis ("smelling" the nutrients) or by elasticotaxis (using the very lay of the land to detect the tension created by the weight of its prey—think of a rock on a tight bedsheet).

The swarm doesn't just bubble over during this hunt, either: There's a lot of coordination going on inside, and how the cells communicate and assign roles to one another is something we're still finding out. We do know that there's a lot of chemical signalling going on in there, and apparently the slime trails have a lot to do with it. Maybe the slime does function as a means of communication, but personally, I'd be surprised if that's the only thing it does.

M. xanthus has one more interesting trait: When the swarm encounters other microbes, it attacks them by spewing forth nasty chemicals and enzymes. *M. xanthus* has hundreds of these, and researchers think we may be able to use some of them as antibiotics against our own disease-causing germs; or we could just let *M. xanthus* loose to eliminate microbes we don't like (such as fungi that rot our crops). You never know when a killer might come in handy.

Jam. Fruit. Martyrs.

M. xanthus has some other surprises up its sleeve: When food is scarce, it has a trick to save itself from starving to death. In areas where population density is high and nutrients are few, aggregates (or "traffic jams") start to appear. Normally, when two *M. xanthus* cells collide, they reverse their engines and move apart, but there are locations at which an *M. xanthus* going one way gets stuck with an *M. xanthus* going another way—and neither can move because they're in front of, or between, or on top of one another.

Sometimes the situation resolves itself: Two cells moving out of the jam in roughly the same direction can meet, signal each other (using a special signal called a C signal, which is reliant on a physical connection between the two), fall in line behind one another, and start moving faster. When that happens a few more times, you get a conga line of *M. xanthus* moving in formation away from the confusion behind them.

If the aggregate is too big, however, these conga lines start moving in circles. The C signals create a positive-feedback loop that creates more and more receptors on each cell surface, and more secretion of the signal, until the level of the C signal reaches a threshold concentration. At this point, up to 100,000 individual cells will organize the aggregate into a structure called a fruiting body. Inside this structure, some of the cells begin to form into spores—dormant, inactive, highly resistant forms that can lie in wait for very long periods until conditions improve outside. Sporulation is a longish process that takes up to three days. Most of the cells, however, form the fruiting-body structure itself. This is a very cooperative endeavor—lethally so: About three-quarters of the participating cells die during the process, probably in order to provide enough nutrients for their brethren to achieve sporulation. Don't forget, this is a period of starvation, so extreme measures are necessary if *M. xanthus* as a community wants to survive.

No one is quite sure yet what advantage this fruiting-body behavior confers on the spores. Why don't the *M. xanthus* cells just form ordinary spores, instead of going to the trouble of constructing a huge, compartmentalized, elaborate structure? It may be that the fruiting body is better for spore dispersal, or that the *M. xanthus* cells, leading such cooperative lives, prefer staying together until times are better, food is plentiful, and they can come out of their shells. I find the second theory more attractive.[10] In such a social species, a cell is bound to want to hang around with its friends, no?

I think the point has been made: To see these microbes only as individuals would be a gross oversimplification—they talk to each other constantly and act together in very complex ways. In a sense, each cell is a participating partner in something that's bigger than the sum of its parts. Should we call it a colony? A community? A superorganism? Labeling becomes tricky in this territory.

Until about twenty years ago, looking at microbes as social organisms wasn't too popular. A microbe was seen primarily as an individual cell, its relationships with its peers confined to straightforward competition. Lately, aided by better technology, observation of organisms like *M. xanthus* has caused scientists to view microbial interactions with more respect. It now appears as if microbes can display a host of quite complex behavior patterns involving communication, cooperation, and deception.

These findings have attracted the attention of evolutionary scientists, who see microbes as having the potential to offer us simple, fast, easily measurable ways to observe and experiment with social behavior. It's surely much more comfortable to poke around in a beaker than to hide in trees for months on end, monitoring warbler mating rituals or baboon aggression patterns.[11] Evolutionary-studies

---

10. Which, of course, counts for precisely nothing, in terms of how true the theory actually is.

11. Misunderstand me not: I'm not belittling macroorganism evolutionary studies by any means. All aspects of nature deserve equally serious attention and good science.

terminology (such as *cheating, kin selection,* and *prisoner's dilemma*) is now beginning to find its way into microbiological literature, and bacterial proteins and plasmids are popping up in social-evolution-speak. The two disciplines are now learning each other's languages; learning to act in concert, to cooperate, to . . . hang on, this sounds familiar. . . .

BONUS TRACK #3

Comma Quick

Imagine you're an *E. coli* bacterium floating around. You're doing fine: The temperature is cozy; the neighborhood's not too crowded; the pH is just so. Maybe you've just multiplied, so you're chatting away with your newly formed double, while munching some nutrients and preparing for the next round of splitting.[12] All is, in fact, well. When suddenly—*Whooooosh! Ktik!*

What was that? It felt like something just whizzed toward you from nowhere and then disappeared.

The good news is you've just had the privilege of witnessing the world's fastest living creature in action. The bad news is it's going to kill you. Bummer. In the precious little time (three to four hours) that you have left to live, you'll probably be anxious to learn more about this fascinating creature—you're doomed anyway, so you may as well broaden your horizons a bit.

You've been hit by *Bdellovibrio bacteriovorus* (which I'll call Bdell, for short). It looks like a very small comma, but it's a predatory bacterium—it eats other bacteria for a living, which is why it's called *bacteriovorus*. It's not the only type of bacteria that does this—we've just heard about *M. xanthus* and their wolf-pack hunting methods, for instance—but Bdell favors a more direct approach: Its preferred method of attack is a solitary torpedo-like charge straight into its prey. It sniffs out prey-rich areas (by chemotaxis) and moves toward them with stunning speed: Bdell has been clocked at 160 micrometers per second (about twenty inches per hour), which might not sound like much to us, but that translates to around a hundred of their body lengths per second. Imagine swimming three Olympic-size pools in a single second, and you'll want to demand some serious steroid testing before they let this guy in the pool.

Bdell achieves these speeds using rotating flagella (the corkscrew-shaped propellers that we met earlier) that chug away at impressive RPMs until Bdell reaches a bacterial cell, which it then latches on to using a hooklike appendage. Bdell then has a few seconds to make up its mind whether it wants to stay or find another cell to invade; after that, the attachment becomes irreversible.

Attack phase over, Bdell now enters cat-burglar mode: The flagellum is discarded, a hole is gently drilled in the prey's outer membrane, and Bdell quietly slithers in. It then seals back its entry hole—not out of altruistic concern for its host's well-being, but to preserve it intact for the next stage.

continues

---

12. "*I'm* the original, *you're* the copy."
"Am not."
"Are too."
"Am not. . . ."

Bacteria usually have an outer and an inner membrane. Bdell nestles in what's called the periplasmic space between the two, and starts really screwing up its victim's life. First, it releases enzymes that degrade the prey's proteins, and then it commandeers the prey's DNA, using the energy and materials from it to replicate inside the prey's body—which has, by now, lost its original shape and turned spherical, because Bdell has eaten away its skeleton.

Now we see why Bdell was so scrupulous about keeping its prey's outer membrane intact: Separated from the outside world by the prey's outer membrane, Bdell doesn't have to share its meal with other creatures, and can take its sweet time devouring and replicating without worrying about any outside interference.

When Bdell senses that it's taken all that its victim can give, all the copies grow flagella and, once more, get into attack mode, exploding the empty shell and swimming like the devil, hungry for the hunt.

Bdell's genome has recently been sequenced, and it turns out to be a rather large and elaborate genome for such a small bug. The extra genes are probably involved in regulating its complicated behavior and allowing it a wider choice of meals: Unlike viruses, which are usually quite specific about which type of cell they can infect, Bdell can feast upon many kinds of bacteria, using different enzymes for different types of prey.[13]

With antibiotics losing their effectiveness nowadays, anything that kills bacteria is bound to arouse some practical interest, and Bdell has become an appealing candidate for antibacterial therapy due to its wide range, its tendency to die out conveniently when there's nothing more to eat, and the fact that it does not pose any risk to humans and animals (indeed, there is evidence that it sometimes lives naturally inside our intestines).

Because Bdell spends its days in close association with the DNA of other cells, one might also suppose that it would be a good subject for studying the transfer of genetic material among foreign cells (a process known as horizontal gene transfer, or HGT); but Bdell actually maintains its genome's integrity remarkably well, and actively resists mixing its genes with any old DNA it finds floating around. This means that it hardly ever mutates in any major way, and we can rely on it to stay as it is and not cause us trouble.

Also encouraging is the finding that Bdell readily attacks biofilms, which, as we mentioned, are getting to be quite a knotty problem in hospitals and many other places. The day we hear a nurse say, "Okay, we should Bdellospray this, just to make sure it's clean," may still be a few years distant, but wouldn't it be great to know that our medicine takes the form of a rushing, ravenous, ravaging, microscopic killer comma?

---

13. The bacterial equivalent, I suppose, of using the correct knife and fork at dinner.

CHAPTER 4

Bugs on the Sly

It takes a bag of tricks to survive and thrive, and a microbe essentially *is* a bag of tricks. In fact, what drew me to microbiology in the first place was the range of things microbes can do—especially the things they can do that we can't, and the things we didn't think could be done at all, until a microbe showed up doing it: deceit, mass poisonings, strange sex, explosives, suicide, security tags, and savage data-processing warfare are all regularly employed in their incessant battle for survival and reproduction.

Blaster Bugs

Try to imagine that you have somehow accidentally swallowed a small but powerful bomb that will go off in your stomach if you move too much. Now hold that thought while I tell you about something called the nitrogen cycle.

The air we breathe is about 20 percent oxygen (which we need), and less than 1 percent of almost everything else (carbon dioxide, smells, pollutants, and other fancy stuff). The rest of it, about four-fifths of all you breathe, is nitrogen. So there's a lot of it about, generally speaking.

Nitrogen, in this atmospheric form, is an (almost) inert gas, which means that it isn't very chemically active. A nitrogen molecule in the air just floats there, minding its own business. Oxygen, on the other hand, is a chemically active gas: It triggers fires and rusts metal, for instance.

When we breathe, our lungs take in oxygen, and our bodies use it, but nitrogen just gets exhaled again—you'd think the human body would use it, but that's hard to do, because nitrogen molecules are made of two nitrogen atoms bound very tightly to each other, and it's difficult to pry them apart. But all living things depend on nitrogen: It's an important part of every DNA molecule, every protein, and just about everything else. So how does it transform into a substance that *can* be used? Bacteria do the job, naturally.

Nitrogen-fixing bacteria have the machinery to convert atmospheric nitrogen into other, more user-friendly, molecules—especially ammonia.[1] For their efforts, they get everything else they need from plant roots, with which they live in the ground, in symbiosis. The plants get their nitrogen in a usable form; we animals get our usable nitrogen from eating plants and other animals; and when we animals are done with the plants and other animals, the nitrogen returns to the ground—either in urine, in other excretions, or as part of an animal's dead body—where is it used again. Everyone is happy. Or dead.

If that were the whole story, though, the nitrogen in the air would have been used up long ago. The reason this doesn't happen is that other types of bugs (denitrifying bacteria) use the nitrogen compounds for their version of breathing, and the result of *that* process is atmospheric nitrogen.[2] Cycle complete.

Or so we thought. About ten years ago, a new type of microbe was found that produces atmospheric nitrogen from ammonia by a rad-

1. These, however, are not the main subjects of this section.

2. Neither are these the main subjects of this section. Patience, people.

ically different method known as anammox (anaerobic ammonium oxidation).[3] This was quite a surprise; for scientists, it felt much like you would feel if you discovered a hitherto-unknown uncle living in your house. Of course, now that anammox bugs have been discovered, microbiologists are finding them everywhere.[4] Turns out they're going to be very useful in all sorts of places, especially in sewage treatment plants, where their ability to convert noxious ammonia gases into innocuous nitrogen will be much appreciated. Anammox bugs don't need oxygen or energy to clean up the ammonia in the sewage, so there are going to be a lot of savings on power and machinery in the new anammox-using plants, which are soon to be taking care of your number ones and twos.

That's nice in a serious, adult way, but not really cool. Here's the cool bit: We're still not sure why, but it seems that for anammox bugs to do their chemical trick, they have to produce hydrazine. Not impressed? Not even when I tell you that hydrazine is the proper name for rocket fuel?

This came as a shock to everyone: Having a factory dealing with highly explosive chemicals near your neighborhood is a serious cause for concern, but having one *inside your body*? Damn.

In addition to being notoriously unstable and dangerous, hydrazine is also highly diffusible and can seep through most regular storage containers. This gives rocket scientists major headaches when they need to figure out how to store it without allowing it to leak away, so it's not surprising that the bacterial way of handling it is also quite tricky: It relies on storing the hydrazine inside a very dense, astonishingly complex internal membrane structure made of carbon

3. This is, finally, the main subject of this section.

4. It's very hard to find something if you don't know what it looks like, what it does, where it is, or if it even exists—especially when it's as small as a microbe.

rings fused together. No one knows yet how the bacteria produce these, but it seems that they do it well, because we don't see them blowing up all the time, or even moving very slowly and carefully, like you would if you had a bomb in your guts.

Clever Cleavers

Bacteria get sick, too, you know, and when a bug sneezes, it explodes. This is the story of what may be the world's smallest wars, and it revolves around a very large group of very small creatures that go by the name of bacteriophages (or phages).

Bacteriophages (which means "bacteria eaters") are viruses that infect—you guessed it—bacteria. Basically, they're like most other viruses: They have an outer shell made of protein, with some genetic material (usually DNA) inside it. As you may have already surmised, they're very, very small.

There are many types of phages, and each type usually infects only one kind of microbe (or maybe a few). Phages have what is probably the highest population count of any type of creature in the world (nobody has actually done an exact head count, naturally). They can be found wherever their hosts are found, which means, as you know by now, everywhere.

A phage that gains entry into a bacterium has two methods of infecting it. In the first, known as the lytic cycle, it takes over the bug's inner processes and starts to churn out hundreds of copies of itself; then it (usually) bursts the poor bug open and begins to spread, seeking more bacterial cells to infect as it goes.

In the second (sneakier) method, known as the lysogenic cycle, the phage integrates its DNA into the bacterial DNA. It does this by cutting the bacterial DNA at a certain point, and inserting its own DNA into the gap. The phage DNA then remains there, getting copied with the bug's DNA as it multiplies, until a certain time (presumably when it deems that conditions are good for the next step) when it switches into the lytic cycle and—*kaboom!*

Bacteria do not take this lying down. They have several defenses against this type of unpleasantness—a sort of miniature immune system, if you will. However, like most security forces, this immune system now faces the problem of recognizing what to attack—how to tell "self" from "invader." The bacterial means of achieving this is through a process called a restriction-modification system (RMS). There are two enzymes involved in this process. The first enzyme tags the entire bacterial DNA (with a very simple chemical group). The second enzyme cuts (or "restricts") any DNA that has not been tagged. In this way, the cell's tagged DNA is safe from destruction; but when a phage comes along, its nontagged DNA is immediately recognized and cleaved to bits. End of story.

Of course, in life, it's never the end of the story. Phages have a bagload of ways to avoid this system: They attack the bacteria's enzymes, so they won't be able to tag or cleave; they tag their own DNA, so it won't be cleaved; they use a different type of genetic material altogether (RNA, or single-stranded DNA), so that the cleavers don't recognize theirs (bacterial DNA, like ours, is double stranded); and so it goes, on and on, measure for countermeasure, throughout the ages.

Scientists love phages for many reasons: They're a useful tool for producing DNA for research purposes, and they offer a very easy way to study viruses and to figure out mathematical models of infection (sort of an "epidemic in a bottle").[5]

---

5. Virologists have it tough: Viruses don't grow outside their hosts, so they need to constantly maintain animals, or animal cells, for the virus to infect. Most of a virologist's time and effort is spent caring for finicky, delicate animal-cell cultures that die if you so much as look at them in a funny way. Bacteriologists, on the other hand, can grow bottlefuls of their specimens overnight, without a care in the world.

One very interesting practice, phage therapy, uses phages to attack bacteria that we don't like—a "my enemy's enemy" approach. It has been studied for many decades in the former Soviet republic of Georgia but, recently, with antibiotics becoming less and less efficient these days, many more researchers are beginning to think seriously about phage therapy as the next big weapon against disease-causing bacteria. They're having some success with it, too; an interesting phage therapy for *Listeria*, the wall-wrecking pathogen, has been developed, for example.

One of the most useful aspects of this whole business, though, comes from the RMS enzymes themselves. In practically every freezer in every lab in the world that deals with DNA are a few dozen tubes of purified RMS enzymes, bought from a supplier. There are hundreds of types to choose from, each isolated from different bacteria, and each cutting DNA in a different, specific place in the sequence. It's amazingly convenient for scientists: They take some DNA, put it in a tube with a bit of enzyme, and go to lunch. When they return, the enzyme has cut every molecule of the DNA exactly where they wanted it to, making manipulation of DNA in the lab infinitely easier and quicker; so everybody's happy.

Except, perhaps, the bacteria themselves. See, the trouble with security measures is that it's very difficult to stop using them. The cleaving enzyme is more stable than the tagging enzyme, which means that it lasts longer in the cell, so if the bug stops producing these two enzymes, the cleaver will hang around after the tagging enzyme has been deactivated and dismantled. The cleaver will then cut nontagged DNA belonging to the bacterium itself, with lethal results for the microbe. So even if there are no phages around, the bacterium is stuck with these enzymes, passing them on as it multiplies. It may have a reduced risk of outside invasion, but the guards are hard to get rid of.

Wriggly Little Computers?

My friend Miki and I were hiking through Tasmania, admiring the view and talking about everything at once, as is our wont. At one point, we started comparing informational systems—I talked DNA and he, a geek extraordinaire, talked IT. We found, ultimately (while absentmindedly slipping in the mud now and again), that just about every natural trick I could mention had its parallel in human-designed computer systems, and vice versa.

It's a curious thing. Computer technology started making its presence felt about the late 1940s. Molecular biology (the confusing name given to the field of biology that deals, hands-on, with the informational content of living things) came to the fore at about the same period, with Watson and Crick's elucidation of DNA structure. In both cases, while there was a history to the developments in these fields—they didn't just spring out of nowhere—the real developmental explosion started around that time, and is still going on today.

I can think of several possible explanations for this fact, starting with pure coincidence and going into more involved explanations, such as the huge overall boost in science and technology that was caused by World War Two. Of course, molecular biology is heavily reliant on computers for processing and analyzing its biological data, so it's natural that the two areas would progress together.[6] Wiser people than I would probably be able to supply even better reasons.

Nevertheless, I find it suspicious. The basic idea behind computer science is sufficiently similar to our grasp of genetic mechanisms that

---

6. A few generations ago, if a researcher wanted to compare two similar DNA sequences, she would have had to painstakingly analyze them, letter by letter. More complex analyses were fiendishly difficult. Today, it takes a mouse click.

we commonly use computer-based metaphors when thinking about genes. I did it myself, in Chapter 1. Could this be a case of analogy abuse? Are we, as a culture, a little infatuated with the parallels between the two fields? Will future thinkers look back on the twentieth and twenty-first centuries and say that the spectacular advances in information technology, as an unfortunate side effect, actually constricted contemporary life scientists' views to some extent? Would we think about DNA in the same way, if computers were not around? I wish I knew. It's tempting to think that we, of the here and now, finally know how nature really works. But past generations thought so, too, with the same amount of conviction—and oh, boy, were they wrong.

Be that as it may, nature does produce an incredible array of marvelous information stunts—so many that I'll have to restrict myself to just a few of the ones that occur in microbes. In order to spark your interest, we shall start off with some sex and death.

Conjugal Matters

Bacteria are famous for not having sex. True, they are well known for interfering mightily and unpleasantly with other species' sex lives, but when it comes to gettin' it on, a bacterium is not exactly a sex machine. Apart from fungi (whose intricate sex lives will not be entered into here, for fear that we will not be able to find our way back out), a bacterium will conduct its reproductive arrangements in a solitary fashion, procreating by dividing and redividing in a way that requires no outside assistance.[7]

However, it appears that this is not the whole story; bacteria have been found to engage in sexual practices, and they are not, I'm sorry

7. Were you expecting some sort of crude joke down here? Shame on you.

to say, of the proper, respectable sort. Microbes tend to have sex in one of three ways: in a distinctly necrophilic fashion; via a third party; or in a way that starts off as intercourse, but ends up being a sex-change operation.

Sex, in its most basic form, can be described as an action that leads to the exchange of genetic material between individuals. In humans, a sperm cell carries one complete set of genes to another complete set in the ovum. Both sets contain unique combinations of gene variants (called alleles), and they join together to produce the complete genome that gives rise to astounding, wondrous creations like you (or to less exhilarating and slightly scruffier creations like me).

In bacteria, the equation normally isn't so elegant. Rather than 1 + 1 = (ultimately) 1, a bacterium can collect and integrate a range of bits and pieces of DNA that it comes across. Ultimately, though, while the systems are radically different, the role of sex in all creatures is the same—to provide much-needed variation that ensures evolutionary flexibility.[8]

An important point should be made here. Evolution is sometimes described as a process that moves slowly from one random mutation to the next, but that is absurd: While that describes the basic mechanism, in reality, once a useful trait or gene evolves in one individual, it can spread like wildfire across organisms and species. A good (and for us, very problematic) example of this process is the way that antibiotic-resistant bacteria have been appearing in hospitals—the genes that confer resistance can be passed on, not only from generation to generation in one bacterial population (biologists refer to this as vertical gene transfer, or VGT), but also between bacteria of different types altogether (horizontal gene transfer, or HGT).

---

8. If you can have some fun in the process, that's a bonus.

Evolution itself is constantly evolving. The methods by which organisms evolve have themselves been getting more elaborate and more advanced over time and, as a result, evolution now proceeds at a quicker and surer pace than it did at the beginning of life.

The occurrence of HGT is also making life harder for microbial taxonomists, the scientists charged with keeping track of the lineage of species (the phylogenetic tree, or the proverbial tree of life). At the microbial level, this tree is rapidly starting to resemble a very tangled bush. When DNA sequences belonging to one species suddenly pop up in another, the very definition of the term *species* gets rather shaky.

Genes that move from one species to another—does that ring a bell? When it happens in nature, we call it HGT; when it's performed artificially, in labs, the term is genetic engineering (GE) or genetic manipulation (GM); in essence, they're the same thing.

Call it sex, GE, GM, or HGT, the most straightforward way for it to happen in microbes is called transformation, and it occurs when a bacterium finds some DNA hanging around outside itself and ingests it. After that happens, the microbe has to make a decision, because DNA (as a physical molecule, rather than an information carrier) is also a handy food source to bacteria. It asks itself, in effect, "Should I integrate this information into the very foundations of my being, or just eat it?" The answer? Different bacteria have different attitudes: Many types won't take any outside DNA at all ("Ugh—who knows where it's been?"); others will only risk it if it's DNA from their own species; and some adventurous types will try anything.[9]

A more orderly though still gruesome route for DNA movement is called transduction, which involves bacteriophages (those viruses

9. This DNA, incidentally, doesn't come from just anywhere: Usually, it's from a bacterium that's just died and disintegrated, which means, in the end, that the DNA gobblers are either looters or necrophiliacs. Fortunately for them, they have no moral code—only a genetic one.

that infect bacteria). As I explained earlier, bacteriophages enter the bacterial cell and take over its DNA-replicating machinery to make more bacteriophage DNA (as all viruses do). They then load that DNA into protein capsules (also unwillingly manufactured by the bacteria) and head off to another cell. It so happens that occasionally the process goes wrong: A bit of the bacteria's own DNA gets packed into the viral capsule by chance, and the capsule then goes on to infect another cell, bringing with it some unexpected genes. Procreation via proxy—there should be a law against that sort of thing, if you ask me.

The final method of DNA movement is the closest to proper sex that bacteria get: It's called conjugation, and this time, it actually involves physical contact between two consenting microbes.

The key player in this maneuver is called an F-plasmid. A plasmid is a circular piece of DNA that is not a part of the main, regular microbial chromosome. There are innumerable types of plasmids (found in many sizes all over the microbial world) that contain all kinds of genes. The F-plasmid mainly contains genes responsible for creating and activating a "sex pilus." A pilus is a hairy, longish appendage that protrudes in a highly suggestive manner from the bug's body (not hard to see why a pilus-wielding cell is referred to, informally, as "male," while a non-pilus-carrying cell is a "female"). Its role, disappointingly, is only to latch on to another passing bug and draw it towards itself. The two cells then touch and engage, and a temporary channel opens up, through which a copy of the F-plasmid is transferred to the recipient cell. Occasionally, some chromosomal DNA is copied and transferred with it, which is the HGT part of it all. The two cells then part company and promise to call the next day, but don't.

All things considered, it's a pretty innocent encounter, don't you think? Allow me, then, to leave you with the following questions:

1. If the plasmid causes all this to happen and ends up being replicated in another cell, can we say that the plasmid itself is a sort of parasite of the bacteria? Is bacterial sex a kind of by-product of an infectious disease? What does this say about nonbacterial sex?
2. We start off with one "male" pilus-producing cell and one "female" non-pilus-producing one. We end up with two "males." Is that, like, beyond kinky, or what?
3. What does the *F* in *F-plasmid* stand for?[10]

That was the sexy part. Hope it didn't ruin your appetite too much. Let us now plunge into the deep abyss of anguish and mortality, and examine what can drive *E. coli* to suicide.

No, No—You're the Disease and I'm the Cure

Let's start with an irrelevancy: Genes have names—usually brief, workaday ones, such as *recA, int, mreB,* or *tetG*. The person who finds the gene gets to name it; however, unlike organisms, genes do not get named after the people who discovered them or the places where they were found. Gene names are, most often, an abbreviation, are meaningless, or have some connection to the gene's role. All in all, they're not hugely creative.[11] As a lover of meaningless tidbits, I was overjoyed to hear the reason behind the *mazE* gene's name: The people who discovered it did not know, at first, what this gene did, so they named it *Ma ze,* which is Hebrew for "What's this?" Later, they

10. I honestly don't know.

11. Except in fruit flies, for some reason, for which gene names such as *smaug, tinman, 18wheeler, technical knockout,* and *cheap date* abound.

found out what it was, but the name stuck, and a related gene was consequently named *mazF*.

This pair of genes has sparked off a rather intense debate that touches on selfishness, death, self-sacrifice, and other fun issues, and that shows no sign of ending anytime soon. It started when *mazE* and *mazF* were found lurking in tandem on a plasmid in *E. coli*. That in itself is not unusual; many microbes contain many types of plasmids, and they have all sorts of genes in them. This, however, was a special case. This plasmid was addictive.

It was found that *mazF* was the gene coding for a protein that was toxic to *E. coli*, and *mazE* was the gene coding for the required antitoxin protein. The two genes are, in other words, responsible for producing a poison and its antidote.

In every living cell, protein molecules are continually being broken down, while new copies are constructed. In some types of protein, the turnover rate is high, and old molecules are quickly replaced by new ones. In more stable types of protein, each individual molecule lasts for quite a bit longer before heading for the cell's recycling machinery. With these two proteins, it was found that the toxin protein is more stable than the antitoxin. So, once an *E. coli* cell received this plasmid (through HGT) and both genes started manufacturing their proteins, the bug was stuck with the plasmid: If it managed to get rid of it, the antitoxin would be degraded pretty rapidly while the toxin lingered on, and the bacterium would be poisoned, leaving the field to only those of its comrades who still carried the plasmid. How cruel is that?

This scenario is a very good example of the renegade-gene mechanism, wherein an individual gene, or a small group of genes, act in order to spread more copies of themselves, with little or no consideration for the well-being of the organism they occupy. Here, one plasmid—two measly genes' worth of DNA—has become a parasite in its own right.

And here, with an elegant little parasitic-addiction module, the matter rested, as far as I knew. As it turns out, during the seven or so years between hearing the story and writing about it, things have been *happening*. Coincidentally, a lot of it was happening in a lab about sixty feet away from the bench where I spent a sizeable chunk of the intervening years. Because labs can be insular places, and research projects demanding, a common malady among my kind is research-student tunnel-vision syndrome: I was so preoccupied with my own project that even the adjoining lab was a hazy blur. I only found out about what was happening recently, reading through research articles on the Internet. I was not too surprised to learn that more addiction modules were found (about a dozen toxin-antitoxin modules have been found, to date, in *E. coli* alone).

Nature rarely contains just a single incidence of a good trick. But in the case of *mazE* and *mazF*, things get even more interesting: It was found that this tag team can also reside inside the *E. coli* chromosome itself. In other words, the addiction module is still a renegade, but now it's dug in deep into a permanent residence. This in itself is not unduly strange. Genes on a plasmid can occasionally get integrated into a chromosome, and vice versa. What is strange is that under extreme starvation conditions, both genes stop producing their proteins, which results in the bug's death.

Let me repeat this, just to let it sink in: *mazE* and *mazF* have collaborated to cause their carrier to commit suicide under starvation conditions. Not exactly what we'd expect from a selfish pair of genes, is it?

What possible benefit does this confer on the genes, or on the bacteria carrying them? Society's to blame, I say. During times of extreme stress, these bacterial communities have found a way to weed out the weakest and least likely to survive, in order to allow the others to live. A starved bug will activate its suicide switch to avoid being a burden on a swiftly shrinking resource stack, and it will even contribute to that stack by becoming food to its friends (bacteria, as we saw, are not fussy about what they eat).

Interestingly, a solitary *mazF* gene waiting for activation was recently found in the genome of *M. xanthus*—the bacteria that lets large numbers of its individual cells die to form spores and a fruiting body during starvation conditions. When hard times strike, the gene will be activated in those individual bacteria that are in the worst situation, to cull their numbers. The weakest link gives way.

Other forms of stress that seem to trigger this suicide response include attacks by phages and antibiotics. The response to phages is thought to be a sort of containment strategy—an infected cell will kill itself to prevent the spread of disease to other cells. This type of response happens in our own bodies, too, when immune cells tell our infected cells to die; or in hospitals, where people with highly contagious diseases are kept in isolation.[12] It's not surprising, then, that bacteria should practice it, too.

Some researchers, though, do not see this response as an altruistic suicide mechanism at all. Instead, they suggest that perhaps we were too hasty in pronouncing death in the first place. Bacteria are so small that it is impossible to determine whether an individual bacterium is dead. Instead, microbiologists use other more roundabout ways to assess whether a cell is alive, such as measuring its growth and activity. They now suspect that sometimes, when outside conditions are very difficult, *mazF* doesn't kill its cell at all; rather, it puts it into a quasi-dormant state—one in which none of its systems is working—and

12. Or in past ages, when people who carried the Black Death were walled in inside their own houses.

the cell hangs in there, inactive and unresponsive (hence dead to all appearances), until conditions improve. This may also explain the module's lack of reaction to antibiotics—quasidormant bacteria are not affected by antibiotics. In other words, it's not dead; it's just resting.

Whatever the overall rationale of toxin-antitoxin modules, they offer us many practical opportunities. Finding a way to convince a microbe that's harmful to humans to do away with itself is very useful—and indeed, a fair number of bacterial baddies are found to contain these modules, so perhaps we could find a way to suggest to them that they'd be better off dead. We could also use addiction modules as safeguards: We could insert them into genetically modified microbes so that the microbe would be reliant on certain conditions and would die off under other conditions. In this way, potentially dangerous microbes could, in effect, be confined to laboratory conditions, which would lessen the chance of one of those engineered-bugs-on-the-loose scenarios that are the cause of such anxiety in certain circles.

Practical issues aside, the argument rages on: Is the "suicide module" a hyper-selfish mechanism or an altruistic one? Does it kill the cell, or just send it into a dormant state? Is it all of these things at once? The latter is frequently found to be the case, but the picture will become clearer, and more elaborate, as research progresses.

In the meantime, this entire section has been something of a downer for me: I thought I was writing about a selfish suicide plasmid. Turns out it's not selfish, and it's not necessarily a plasmid; and now, it seems it may not even trigger suicide, after all. Researchers now understand this plasmid better, but we've come right back to the question, *Ma ze?* All this to give you the freshest, most up-to-date version of things. I hope you're happy (*sniff*).

The Sigma Factor[13]

For a microbe, life is a constant battle to thrive within its surroundings. Outside conditions can, and do, change rapidly (and sometimes dramatically), so microbes must be very adept not only at surviving in a certain environment but also at responding to these changes with speed and precision. They are helped, in this regard, by sigma factors.

Sigma factors are a group of clever little decoders, dictionaries, or encryption keys that exist in many types of microbes, in their RNA polymerase. RNA polymerase, you'll remember, is an enzyme that recognizes the exact place in an organism where a particular gene starts. These places are specialized spots on the DNA sequence that are called promoters. The enzyme attaches to the DNA at that promoter, unzips it (so that a single strand can be read from it), and makes (or "transcribes") an RNA copy of the DNA sequence, like a photocopied page.

RNA polymerase is composed of six subunits, and one of them, the unit called sigma, has the job of recognizing the promoters. The sigma finds the promoter and attaches itself (along with the rest of the RNA polymerase) to the DNA. Once the RNA polymerase is steadily transcribing away, the sigma's job is done, and it drops off.

It turns out that microbes have more than one kind of sigma factor: *E. coli,* for example, sports no less than eight different types. Under certain conditions, in a classic case of "cometh the hour, cometh the factor," the regular sigma's role is taken over by one of the alternative sigmas. When the bacterium is exposed to uncomfortably

---

13. This is probably the most straightforward heading in this book: Surprisingly, this section is actually about sigma factors. I've been trying for titles that sound intriguing and alluring, but this one already sounds like a paperback thriller.

high temperatures, for instance, the regular sigma is replaced by a specialized "heat shock" sigma. This alternative sigma recognizes a different type of promoter—the kind located at the start of genes for "heat shock" proteins that, in one way or another, help the bug cope with the heat. Other sigma factors are put into action when the bacterium is starved for nitrogen, or needs more iron, and so on.

This is a rather neat arrangement. The bunch of genes that need to be activated at a certain time can be dispersed all over the chromosome, and will be called into service by their own unique master key.

But how does a cell know when it is time to switch sigmas? Ultimately, this is one of those origin questions, like, "Who checks the dictionary for spelling errors?"[14] The original stimulus is usually a physical one; for example, in heat-shock situations, it is the heat itself that causes a certain protein to come apart. That protein is normally tied up to the heat-shock sigma factor, so that when it is gone, the sigma factor is able to commence work.

There are, of course, many other mechanisms at work that fine-tune the entire microbe reaction (efficient regulation is the key to success in the cellular world, after all), but the low cost and high benefit of the sigma-factor system makes it a winner all around.

Responsiveness and flexibility are highly advantageous properties for a microbe. There is, however, another way to approach the task of survival. It is not the most innovative approach, admittedly, but sometimes it seems to work.

---

14. This question bothered me a bit when I was younger (I was that kind of child). I finally laid the matter to rest when I learned that Shakespeare even spelled his own name in several different ways. If he wasn't bothered with correct spelling, why should I be? Alas, this argument failed to carry with my teachers.

Simply the Best

The best criminal, it is said, is one who never gets caught, is never suspected, and is never even heard of. I always imagine a quiet, regular-looking guy who lives happily with his wife and kids on an island paradise, and earns some modest millions by channeling tiny, unnoticed sums into his account, every day, from every single account in a large bank.[15] We'll never know if he exists. That's his genius.

The world's most successful microbe is much the same, really: You've never heard of it; it won't make you ill; it isn't spectacular or very interesting; and we don't even know what it looks like. It lives in the sea, off nothing much, scraping an existence from dead organic matter that has dissolved in the seawater. It didn't even have its proper name, *Pelagibacter ubique* (*P. ubique*), until quite recently, having to contend with the dull designation SAR11. Its numbers are just about the only remarkable thing about it: There are an estimated 10^{28} *P. ubique* bacteria in the world—that's about 30 percent of all living things. Think about it: Out of every ten organisms—beetles, crickets, Belgians, bats, mosquitoes, flies, germs—three are *P. ubique.* Let's hope they never learn to text message, or reality TV will become even weirder.

What makes *P. ubique* such a successful reproducer? The brief answer is its streamlining capabilities.[16] In addition to being quite small, *P. ubique* has no dead weight in its genome: It has 1,354 genes (humans have about 30,000, and *E. coli* 3,000, as a comparison) and nothing else. Many other organisms harbor some extra DNA in their

15. His features, funnily enough, resemble those of yours truly.
16. The long answer is, well, long.

genomes—mostly genes for proteins that are activated only in certain conditions, occasionally an unused copy of a gene, a little ancient viral DNA left over—like old spare parts in a mechanic's workshop. *P. ubique* has none of that. There are microbes with smaller genomes, but they're all parasites that rely on their host for many of their functions. *P. ubique* needs no one but itself: With its minimal genome, it can reproduce quickly and efficiently, which it does.

Extra genes, as I've explained, are not necessarily a bad thing: They provide flexibility when conditions change. For instance, if a microbe's environment suddenly runs out of one kind of foodstuff, a set of genes that enable it to eat another kind of foodstuff is very helpful. *P. ubique* demonstrates the opposite approach: Even if there is suddenly heaps more food available, it can't use the food to grow any quicker. If it runs out of its food, it starves. It's a one-trick microbe that bets its existence on the ocean not changing too much. For the last billion years or so, it seems it has kept ahead of the game.

Darwin to Cairns, and Back

The term *Lamarckism* is a no-no in modern biology. In the nineteenth century, it was a popular theory of evolution, but then Darwinism showed up, making much more sense and, more important, proving itself to be sound. Jean-Baptiste Lamarck's theory, on the other hand, is still ridiculed to an extent, and is often used in evolution textbooks to contrast with our understanding of the way that nature actually works. It states that an organism passes on to its offspring characteristics that it has acquired during its lifetime—so the giraffe's neck begins to get longer as every generation strains for ever-higher leaves, or the blacksmith's sons begin to be born with the potential for more muscular arms, for example.

In some respects, the ridicule is unfair: The theory was as logically sound as any rival theories that were around in the early 1800s. Also, teachers of evolution tend to focus on the most erroneous elements of Lamarck's theories, presenting them as if they amounted to his whole argument.

But all that is only of interest in a historical sense. A much more startling proposition was put forward in 1988 by John Cairns, a Harvard professor who found evidence that *E. coli* were behaving in ways that appeared to follow Lamarckian mechanisms. This was treated with much suspicion. Darwinism was (and still is) a very robust theory, and previous announcements of the return of Lamarckism had been fraught with political motives: The Lamarckist agricultural theories of Trofim Lysenko that were embraced in the U.S.S.R. in the mid-twentieth century, for example, caused much poverty and starvation, until policy makers finally abandoned the conviction that plants were a flowering type of Marxist.

Nevertheless, here was Cairns, with an indisputably scientific experiment that showed *E. coli* to be evolving much more quickly than the random mechanisms predicted by Darwinism would have it doing: A Petri dish that contained virtually no energy source besides lactose was spread with *E. coli* bacteria that had a defect which affected their lactose-consumption abilities. The defect was carefully induced so that a single mutation at the right spot would fix it. This was a case of mutate or die: Any individual bug that did not mutate its lactose-using gene back into activity would leave its descendants to starve to death within a few short generations. The rare revertants—individual bacterial cells that had reverted to an activity they once had, but had now lost—would be the only ones to see the day through.

The researchers expected, based on an established rate of mutation, that a very small number of bacteria would randomly and serendipitously revert, before the rest of the food was gone, to become lactose users. But the number of actual revertants was much, much higher—about a hundred to a thousand times more than expected. Could the bacteria, in some way, sense that lactose was around and fix their lactose-utilizing genes? Could a living thing actively change its DNA to suit outside conditions?

To say such a thing would be to put into doubt one of the two basic principles of Darwinism; it would be much like suggesting that people could somehow will themselves taller. While the scientific community is relatively easy on revolutionary ideas (and, in the ideal, encourages them), casting aside a theory that had been working so well for so long is not done lightly, especially if the results seem to suggest that bacteria can do something as audaciously advanced as planning ahead.

Cairns's results triggered a good crop of experimentation and discussion. Within a few years, the idea that bacteria had the capacity to plan was dismissed, on the grounds that it did not have any evidence to back it, and the ghost of Lamarck, tentatively resurrected for a short while, was laid back to rest—although recently some of his ideas are being reconsidered by modern scientists, in one of biology's most interesting current debates. But that's a story all to itself.

The two rival theories currently attempting to explain this phenomenon are squarely back in Darwinian territory.[17] The first one is highly technical and will not be explained here; the second posits that, under extremely difficult conditions, a bacterium may throw caution to the wind and allow its DNA to become less tightly regulated, so that error-fixing mechanisms are switched off, and the mutation rate hits the roof. According to that theory, this is why

17. With each theory's supporters referring to the other in the short, cold, formal manner that is the scientific version of poisonous invective.

lactose-using revertants suddenly popped up in Cairns's starved *E. coli* Petri dish.[18] This is definitely a kill-or-cure solution; living things, as a rule, do not like mutations messing up their DNA. There are entire regulatory systems in operation whose job it is to prevent precisely this sort of thing (remember D-Rad's toroids and compartmentalizations?). Nevertheless, the entire history of life has always been a delicate balance between stability and change, and it is reasonable to assume that this balance is liable to shift, if necessary.

If this theory is true, it's potentially very good news: If we're able to pinpoint the triggers for increasing mutation rates, we could design "evolution stopper" drugs to target them. Taken jointly with a new antibiotic, these drugs could prevent bacteria from developing resistance to the antibiotic. Research into cancer (which is essentially a cell breaking out from regulatory mechanisms and accumulating mutations that are beneficial to itself, with disastrous results for the rest of the body) could also benefit enormously from advances in this area.

With all these shiny new theories and evidence, it's worth remembering that the idea that the environment may increase variability among species is not new; good old Darwin had suggested that this could be the case 150 years ago.

Masters of Random Disguise

The *Mycoplasma* family are an intriguing group of microbes. They're a favorite of mine, though I'd probably change my mind if they infected me—but even then, I'd have to admire them for fooling most of their adversaries, despite being so small.

---

18. Why did the lactose-unrelated mutations not show up? Because they had accumulated lethal mutations and died.

Aside from viruses, *Mycoplasma* are the smallest living things on Earth. The smallest of them all, *Mycoplasma genitalium* (*M. genitalium*), is roughly ten times smaller than the average bacterium and contains one-tenth of the DNA.

This raises an interesting question: How do *Mycoplasma* manage to get along with so little genetic information when all the other microorganisms need more genes? Do other bacteria have a whole lot of superfluous genes hanging around?

As far as we know, the answer to the second question is "Usually not." DNA that appears to serve no purpose can be found in eukaryotes (such as humans), although this lack of purpose, too, is far from certain, and recent evidence suggests that there may be some interesting surprises hiding in this "junk DNA" of ours. On the other hand, microbes, as a rule, are very efficient, and evolution has streamlined their genetic content to a high, though not always perfect, degree.

The answer to the first question is that, unlike other microorganisms, *Mycoplasma* outsource: They're obligatory parasites, which means they live inside a host, relying on it to provide them with all sorts of necessary stuff that other microbes usually have to manufacture for themselves, such as nucleic acids and nutrients. They even make do without a cell wall, which is found in almost all other microbes, trusting their environment to be stable.

Without a host—such as a cow, sheep, chicken, or a human—*Mycoplasma* simply cannot live and prosper in nature. But hosts don't like to be taken advantage of, and their immune systems are primed to locate and destroy these freeloaders.

If the hosts succeeded, that would be the end of the story, but this germ employs a sneaky trick called antigenic variation that saves its skin. It works like this: *Mycoplasma* have a set of proteins displayed on their cell surface that the host's immune system recognizes. Only one of these proteins is displayed at a time but, once in a while, randomly, a single germ will change its displayed protein. When the host's immune system sets out to find the germs, it doesn't recognize the germs that have the new protein. It annihilates the original protein-displaying germs, but the new protein-displaying germs are free to go forth and multiply. As the immune system reactivates to recognize the new protein-displaying germs, the germ generates yet another, different, protein; and round and round it goes.

In the host, this cyclic pattern manifests itself as chronic illness, with symptoms that come and go: When the *Mycoplasma* multiply and the immune system is active, the signs of battle manifest as inflammation; the immune system then triumphs temporarily, the *Mycoplasma* are almost destroyed and inflammation recedes; then the new protein-displaying germs begin to multiply, and we're back where we started from.

Breathing Is Overrated

Oxygen—what a silly idea: It's highly unstable, it corrodes metal, and it's dangerously flammable. It's also very addictive. Try giving it up—it's impossible. Truly nasty stuff.

I admit that this is not the usual view: Most people think of oxygen as the most basic, obvious requirement for life—and for us, it is; but this wasn't always the case. Oxygen, in the form we breathe it, is basically toxic waste.

Back when there was no life on Earth, there was also very little oxygen. The trouble started when early microorganisms, billions of years ago, developed a way of using solar energy for their own benefit. This process, known as photosynthesis, was very efficient indeed, but it had an unfortunate side effect: It polluted the atmosphere with oxygen. In those days, there was no government to regulate emission levels, and this irresponsible practice was allowed to continue, unchecked, for millions and millions of years until the whole atmosphere was drenched with the stuff. To our shame, these photosynthetic bacteria and their multicelled descendants (plants) continue to spew forth oxygen fumes, even today, despite the valiant efforts that arsonists, logging companies, and strip-mining ventures are making to reduce their numbers.

The effects of this ecological catastrophe were huge; today, we have only a vague notion of the vast number of microbial species that were driven to extinction, poisoned by oxygen. There were survivors, though.

A substantial fraction of microbes live, to this day, as strict anaerobes—they cannot tolerate even a mere whiff of oxygen. We find the notion of creatures that are poisoned by air curious, even ridiculous, but they exist—often in quite surprising places, like our own gums.

The rest of us are microbes (and descendants of microbes) that, over the millennia, adapted to the rising levels of oxygen by learning to use a harmful and dangerous substance.

That we now take oxygen use for granted should not lessen our respect for this remarkable feat of adaptation. Just to make the point: The chemical reaction that uses oxygen to provide our bodies with energy is a tightly regulated form of combustion, the same reaction that powers the internal-combustion motors in our cars (though with different substrates)—the same reaction that produces fire.

But how did those ancient microbes manage to survive? All living things are shaped by their environment: They evolve to tolerate what is available in their vicinity, and then use it. What they lack will either be sourced from somewhere else, or they will produce it themselves. This is, for instance, the whole idea of vitamins—a vitamin is just a word for a substance that the human body needs but isn't able to produce independently. The human body manufactures thousands of different substances that it requires, but there are a few dozen substances that humans consistently find in our regular food sources, so we evolved to stop producing them. While other organisms go through the trouble of producing vitamin C for themselves, for example, we simply eat those organisms; it's easier. If the oxygen levels rise, he who finds a way to live with it survives. He who finds a way to *use* it has an advantage over those who merely tolerate it; and so the game continues.

The ability to use oxygen is a mixed blessing, though, and I'll try to explain why by using a sports analogy. I once read a comment by a sports-medicine expert who said, "A world-class professional runner is, by definition, an injured person." This puzzled me: Professional athletes seem to be very fit people, what with the muscles, and the fast running, and all that. Surely they take good care of themselves? But the expert explained that because athletes are always striving to be at the very peak of their abilities, they are living on the edge; a bit too much effort, and they're in injury territory. Those who injure themselves are, of course, out of the running, but so are those who don't try as hard. In the competitive world of athletics (in which superb natural talent is only the entry requirement), the one who wins is the one who has managed to get as close as possible to his or her edge, without stepping over it. So it is in the race of life. Different organisms and species are in everlasting competition with each other.

An organism using oxygen can extract nineteen times more energy from its food than an organism that doesn't use it. Without this added boost to their energetic balance, our ancient microbial forefathers probably would not have been able to develop into advanced, multicelled creatures in the first place.

But this huge advantage over non-oxygen-using organisms carries its own price. Oxygen's high reactivity is due to its ability to create free radicals: highly reactive, though short-lived, molecules that cause cell damage. Because these free radicals can also attack DNA to create harmful mutations, oxygen is, ironically or not, a powerful cancer-inducing agent. This is why we are encouraged to take antioxidants—substances that prevent free radicals from harming the body's tissues.

Some types of organisms occupy the middle ground between the two extremes of relying on oxygen and dying because of it. Aerotolerant creatures (such as the lactic-acid bacteria found in yogurt) don't care one way or the other about oxygen, while facultative anaerobes prefer using oxygen, but can live without it. Yeast is one example of a facultative anaerobe, which is why we seal flasks when making wine: If a yeast cell has no oxygen to use, it will revert to non-oxygen-using mode (fermentation mode, which produces alcohol as a waste product).[19] Once oxygen is available, the yeast immediately switches over to the more energy-efficient respiration process, whose waste product is not alcohol, but acetic acid—vinegar.

The rest of us are stuck with oxygen. Creatures like us are called strict aerobes, and we must provide energy for ourselves by means of barely controlled fires inside our bodies. We are destined to be playing with fire for all time.

---

19. Alcohol a waste product! We shouldn't be surprised: Alcohol is detrimental to humans, too; it's just that we enjoy poisoning ourselves. Sad it may be, but a fine Scottish single-malt whiskey is, at the heart of it, just yeast excrement.

BONUS TRACK #4

The World's First Microbiological Sketch

Dud: Nothing like a cup of tea, eh, Pete?

Pete: Nothing like a cup of tea, Dud.

Dud: O'course, a half a cup of tea's a bit like a cup of tea.

Pete: Only half as nice.

Dud: Yes, half as nice. But twice as nice as an empty cup, mind you.

Pete: Yes, you're right. It depends on your point of view, really.

Dud: Yes . . . I'm awfully sorry about the bread and butter—I thought I had plenty, but the bread's gone moldy all of a sudden.

Pete: You've got to keep an eye out for mold. It's a kind of fungus, y'see, and they can spring some surprises on you. Matter of fact, the other day, I was reading about this fungus that eats nematodes.

Dud: Nematode . . . that's one of them little green frogs, right?

Pete: No, Dud, not a frog. A nematode's a kind of worm, only very tiny; so tiny that you can't really see it.

Dud: So how can you eat it, if you can't see it? You'd be trying to stick your fork in it, going, "I wonder if I've got it this time. . . . No, hang on, missed it again." You'd never finish your dinner.

Pete: Well, you can't eat nematodes, because you're not a fungus—though you look a bit like one. Funguses can eat small nematodes, bein' likewise very small.

Dud: How does it catch 'em, then? Does it jump on 'em from behind and grab 'em?

Pete: No, a fungus can't do that, what with having no legs. It ambushes the nematodes, cunning like, with its clever traps.

Dud: Doesn't sound like much of a life, I'll tell you that. If I was sittin' up there in heaven and they came 'round to tell me that I'm going to be reincarcerated as a fungus, I'd demand to see the management.

Pete: I think you'll find that it's *reincarnated.* Incarcerated's when they chuck you into jail and you're stuck in a cell.

Dud: I don't see much of a difference between bein' stuck in a cell or bein' a fungus. I wouldn't wanna spend my life fungusin' around, eatin' raw worms.

Pete: Well, maybe it's a good life when you get to know it. If you'd go and ask a fungus, perhaps it wouldn't like to be a Dud, sitting around drinking tea. Maybe it likes raw worms. "Nothin' better than a nice spot o' worm for lunch," is prob'ly the fungus's opinion on the matter.

continues

Dud: Depends on your point of view, really.

Pete: Yes, well, I'm not sure funguses have points of view. Not very strong in the brain department, if you know what I mean.

Dud: So how does it catch those worms, then?

Pete: Well, you got three sorts of nematode-trappin' fungus: the sticky sort, the tangly sort, and the constricting sort.

Dud: What, does it sign the worms up for the army?

Pete: No, you're thinking of conscripting. Constricting's when something tightens up on you.

Dud: Oh, right.

Pete: In any case, the sticky sort has these sticky substances stickin' out, which the nematode gets stuck on. The tangly sort has a sort of net, or a very small maze, which the nematode gets tangled in and can't break free from. The constricting sort we were discussing just now is even cleverer: The fungus has a sort of ring, made up of three special cells in a circle, hangin' from its side, quite big enough for a nematode to pass through; but if one does, and if it slightly touches one cell, then all the cells together will go *whooomph*, inflatin' very quickly, like those airbags you get in cars nowadays, and that traps the worm inside the ring, so it can't break free.

Dud: Catching it 'round the neck?

Pete: A worm doesn't have a neck, Dud. Or, rather, it's got nothing but neck. Depends on your point of view, really.

Dud: So what does the worm do, then?

Pete: I expect it wriggles a bit, thinking, *Should've seen that one coming. What a fix.* Then, after a while, it dies, and the fungus eats it. Grows into it, actually.

Dud: At least it waits until the poor worm snuffs it. Not everybody's that decent in this day and age.

Pete: True, very true.

Dud: Where'd you read all this stuff, anyway?

Pete: Well, just the other day, this guy comes up to me in the street and says, "You're Pete, right? Thought I recognized you." Never saw him before in my life. So I ask him, "Where from?" and he says, "From this book I'm reading, *The Invisible Kingdom*. You're in there carryin' on about worms." And I think, *Funny, you know, I don't recall doing that, 'specially not in a book.* But he shows it to me, and there I am, discussin' this fungus with you in a very knowledgeable manner. Quite interestin', really.

Dud: What's the book like, then?

Pete: Not really my cup of tea.

CHAPTER 5

Bugs on Us

Imagine that a friend comes up to you and says, "Have you heard? Every microbe on Earth has decided to leave. They've all gone!" Had that happened before you wisely decided to read this book, you might have shrugged, said "So?" and gone back to whatever you'd been doing. Had you been slightly more knowledgeable, you might have recalled all of the horrible diseases that microbes cause and said, "Good riddance." If, however, you've been paying attention up till now, you'd undoubtedly realize the full implications of this new situation:

1. Life on Earth won't be remotely the same, ever again.
2. We'll all be dead within days, if not hours.
3. There's a lot of stuff around the house that we won't need anymore.

The first two realizations are depressing, so let's concentrate on the third, which will offer us the chance to sell things, make money, clear up some space at home, and generally make life more enjoyable. Turning our mind's eye homewards, then, if there are no microbes anymore, what don't we need?

Well, most of the big white boxes, for starters. If there are no microbes, food won't spoil, so you don't really need a fridge. Milk, eggs, and meat will all be perfectly fine in the cupboard. You may want a small bar fridge for keeping drinks and ice cream cold in summer, but that's it. So dump the fridge.

Most of the pantry can go, too: Out go the tins (why bother?), out go the wine racks (no wine without yeast; no beer, either, for that matter), and out goes the water purifier.

Next, the dishwasher: You really don't need to go through all that just to clean some dishes. Give them a quick rinse in the sink, and that's it. Food scraps will no longer make you ill or taste bad, so you can leave them on, if you like.

The same goes for the washer and dryer. Who cares if your clothes have been worn? If they haven't got big, ugly stains on them, you can just go on wearing them. They won't smell, don't worry: It's the microbes on your skin that make body odor (well, used to). Now they're gone—so no washing necessary. If you're like me, this will mean you can wear the same shirt every day, until it disintegrates completely.[1] Even those of us who enjoy clothes and fashion could probably free up some closet space—it would certainly come in handy for storing the surplus milk, if the cupboard is full.

So out with washer, dryer, clothesline, detergents, deodorant, soap, and the entire shower, if you want. No need for brushing teeth anymore, so chuck away the toothbrush, paste, floss, and mouthwash, as well as the antiseptic and bandages—all into the trash can. We

---

1. I had one T-shirt that I wore relentlessly as a teenager. It was not a thing of beauty, plus it was already twenty years old when I inherited it. Eventually, I couldn't tell which were the original holes meant for putting limbs through. I had to rescue it twice from the trash, and once from my mother's attempt to use it as a rag. Finally, my sister, in desperation, had it framed, and presented it to me in a short but emotional ceremony.

don't really need a toilet anymore, so a lot of expensive plumbing is now unnecessary . . . on we go—and that's just the household stuff.

When you look at it this way, a huge portion of the march of human civilization can be seen as a series of attempts to ward off microbial influences, while an additional portion concerns the attempt to use them (which will be the focus of the next chapter). We can't review the entire range of human-microbe interactions here, of course, so let's stick to the important, interesting, or especially uncommon ones.

First, though, allow me to talk about one of the most common things around.

The Common Not-So-Good

My father-in-law, God bless him, has a cold remedy he swears by. The details are something of a family secret, but I am allowed to disclose that it involves boiling Coke and a fair measure of garlic. The singular experience of imbibing this potion I leave to your imagination; suffice to say that it will cause anyone who drinks it to wish much more fervently for scientists to hurry up already and find a cure for the cold.

It seems like such a small thing to ask, doesn't it? Medical science has triumphed over such horrible diseases that curing the sniffles shouldn't be so hard—a matter of weeks, if one puts one's mind to it, surely?

Alas, that is not the way things work. A major problem is that the common cold is not really a disease. It is a general name for a collection of symptoms that can be caused by over a hundred types of viruses from several different families. We didn't even know what a virus was until a few decades ago, but have been wiping our runny noses since prehistoric times.

When it comes to treating colds at their core, there's still a lot we don't know, because there isn't one core at all—rather, there are dozens of different ones. What's more, these viruses can mutate very quickly, so that even if a solution *is* found for one type of cold-causing virus, it may be useless just a short while later. Our immune system has the same problem, which is why we suffer colds about twice a year, instead of catching a cold once and staying naturally immunized against it for the rest of our lives, as we do for proper diseases.

All this makes a cold a very difficult condition to cure: We treat the symptoms, most of which are caused by the body's struggle against the infecting virus, and we trust the body to do the rest of it.

Breaking news: In February 2009 a team of researchers announced that they had mapped the DNA sequence of all ninety-nine known variants of rhinovirus, a family of viruses responsible for most cases of colds worldwide. This is a major step forward and will help us figure out what it is exactly that these viruses do and how to stop them from doing it, but effective drugs, vaccines, and eradication are still some way off. So in the meantime, though I'm no medical authority, I feel you're entitled to the latest in expert opinion on the treatment of colds. I therefore leave you with some practical advice.

To prevent: wash hands.

To cure: have a rest and some chicken soup.

Avoid: boiled garlic Coke.

Remix

The flu is different from a cold. They are both viral diseases, but a different family of viruses causes them, and they work in different ways; however, they do share similar symptoms and a tenacity that's hard to beat.

Both seem to make regular appearances in winter, for one thing, and this may be due to a number of factors. In winter, there is closer contact between people (because we spend more time indoors); colder temperatures and drier air contribute to the lowering of our immune resistance and allow viruses to survive longer on outside surfaces; and less sunshine means that we tend to lack vitamin D.

We don't march to the doctor to get our cold vaccinations, because no such thing exists (for the same reasons that our attempts at cold cures have been hindered); however, fortunately, flu vaccinations are available, and we can (indeed *should*) receive them once a year. That way we can avoid the annoying sympt. . . . Hey, hang on a minute. Why once a year? Most vaccines we receive are administered to us once in our lifetimes, with perhaps a booster shot or two a few months later. How is influenza different?

It's different because the influenza virus can do an odd thing or two with its genes. The first is neat, but familiar: A simple mutation mechanism in the virus's genes causes the surface proteins of the flu virus to mutate, so that their surface appearance changes and our immune system cannot recognize the new strain.

This is normal, unexceptional viral epidemiology; however, once a decade or so, something very different happens, which accounts for the bigger, more troublesome flu epidemics. To understand why this happens, we need to understand viruses a bit better. A virus is a very different thing from a bacterium: While a bacterium infecting a cell keeps itself together and multiplies, a virus disassembles itself inside the host cell. In many cases, the outer structure of the virus doesn't even enter the host cell; it just injects its inner contents (DNA, or RNA, and sometimes a few proteins) into the cell. These intruding components take over the host cell's manufacturing mechanisms (which have, up until now, been taking care of the cell's regular upkeep), and force them to manufacture numerous copies of virus parts instead. Then the new viruses reassemble and head off in search of new cells to infect.

A virus's genetic material is usually just one string of DNA or RNA, which is very convenient for quick replication. A flu virus, however, is different: Its genome is made up of eight short, separate segments of RNA. This is, in nearly all cases, pretty irrelevant for us; the virus does what viruses usually do, only it does the RNA replication thing eight short times, instead of one longer one. The flu virus has ways of ensuring that, most of the time, these eight segments all go into one outer shell to make a new, working virus particle. Because the genetic material is RNA rather than DNA, it mutates at a faster rate, meaning that, on average, each single virus is slightly different (which accounts for the year-to-year variations in strains).

But that's not the big problem. The big problem comes along very rarely, when a single person or animal (usually a pig) is infected with two different strains of the flu. If that happens, it may happen that a single host cell will be attacked simultaneously by two versions of the flu virus. Let's say that a pig happens to catch one strain from a duck and a second strain from a person (if you don't see how this could possibly happen, you need to get out more), and both strains then infect cells in the pig's respiratory tract. By chance, virus particles from both strains have reached and infected a certain cell at the same time, and now we have the eight RNA segments floating around in the cell and being packaged into new virus particles. Instead of 1 2 3 4 . . . , we now have 1a 1b 2a 2b 3a . . . , and there's a good chance that some particles will contain segments from both versions—for instance, 1a 2a 3b 4a. . . .

Voilà—we have just witnessed a reshuffling of the cards: A new combination has been created, and a new flu-virus strain is unleashed unto the world. If the reshuffling influences just the inner workings of the virus particle, it's no big deal to us; but if the surface components of the viral cell are reshuffled, it means that the new virus will present an enigma to our immune system, making us susceptible to infection.

The virus can then spread to other cells in the pig, to other pigs, back to ducks and humans, and from humans to humans, until the hills are alive with the sound of coughing.

You'll notice there were a lot of *ifs* in this description: if a pig catches two strains, if it happens in one cell, if the particles are reshuffled, if it's the outside proteins that get remixed. There are more, still: if the new combination is still infective from cell to cell, if it's infective from pig to pig, if it's infective from pig to human, if it's infective from human to human . . . what are the chances of all that happening? Not large, but if you take a very small chance and repeat the event many, many times (do you realize how many flu-virus particles there are knocking about in one body, let alone the entire world?), occasionally even small chances materialize; and it only needs to happen once, anywhere, for the entire world to be affected. That's the power of mutation and natural selection.

Reshufflings are very uncommon among viruses (as far as we know), but they sound oddly familiar.[2] Although the details are very different, reshuffling of genetic elements is precisely the point of sexual reproduction—it happens every time a new human is created. As an individual, you are, a virus would say, a new substrain of humanity.

The avian flu strain, back in 2005, provides a good case study for all this. It was a virus that got stuck at the human-to-human barrier.

---

2. Interestingly—*very* interestingly—the 2009 study I mentioned in the previous section shows that cold viruses can perform a version of the same mixing trick as flu viruses, the little rascals. Watch this space for developments.

While humans have caught it (256 deaths were recorded by mid-March 2009), they've caught it from birds, not from other humans. This is good news (unless you're a chicken): Although the death of any number of people is never a good thing, what could potentially happen is much, much worse. The Spanish flu strain of 1918, for example, decimated somewhere between 2 to 5 percent of the entire human population. We hope that we're better prepared now than we were over ninety years ago; we have better treatments for the flu itself, and for its symptoms and complications; we know more about viruses; and we're developing vaccines against possible future outbreaks. It may also be that in human-to-human form, it won't be as bad as we fear; nevertheless, we're not safe yet. If truth be told, we never are.

Pray, do not be disheartened—I will now attempt to explain, in brief, nearly the entire history of the world, and why we're not all dead by now. I'll also mention frogs.

Jumping the Fence

When I was studying microbiology, I found myself part of a team researching *Mycoplasma* (the "masters of random disguise" from the last chapter). I was concentrating on the genetic elements in *Mycoplasma bovis,* which, as the name suggests, is found in cows. Other types of *Mycoplasma* studied in our lab included *Mycoplasma gallisepticum,* found in chickens, and *Mycoplasma capricolum,* found in goats.

I was several months into my project, busily mapping out the DNA sequences of my bovine pathogen and comparing it with sequences from *Mycoplasma* species in chickens and humans—to map out the relationships between the genes, the pathogens, and the hosts—when I came across the book *Guns, Germs, and Steel* by Jared Diamond. Immediately, I felt like kicking myself. I'd been so caught up in the details of my project that I had failed to ask one obvious question—the very question that Diamond had answered in his book. Why *should* a human pathogen be related to the pathogens of cows or bloomin' chickens?

Think about it: As a species goes through evolution, so do its parasites and pathogens. Because their ecological niche is another organism, it's only natural that at least some of their changes are reflections of, and adaptations to, the changes that their host goes through. If a host species changes its diet, or habits, or surroundings, the parasite must adapt, or it'll eventually die off. If we look at two closely related species of animals, we can reasonably expect that their parasites and pathogens will also be similar and related, at least to a degree—and, conversely, that the further away they are on the phylogenetic tree, the more different their pathogens.

How, then, do chickens relate to our evolutionary heritage? Perish all images of naughtiness from your mind.[3] The answer, stripped down, is that humans and livestock have lived in close physical proximity for thousands of years and, during that time, we have occasionally caught something from them. Such an event is no small feat. As we saw with the flu virus, for a microbe to jump between species is rather rare, and the further apart the species are on the phylogenetic tree, the smaller the chances of this happening—you cannot, for instance, expect to catch frog diseases, no matter what you do with the frog. This is because the same individual cell that was infecting species X must now not only survive in species Y (which is, as

3. Unless you find that you enjoy them.

we said, a different environment), but must also be able to be passed on between members of species Y. That's rather a tall order for evolutionary processes. Nevertheless, constant contact throughout the ages brought on the occasional interspecies fence jump. After an initial period of acclimation, the new host species would become able to handle the newcomer with some success and not die off en masse.

This brings us to plagues—the most horrific result of human–microbe relations. Various plagues that have been caused by microbes have exercised their dreadful toll upon humanity throughout history. It is only recently, with the aid of vaccinations and antibiotics, that we have managed to protect ourselves, to some extent, from their damage. Yet even at the worst of times, a plague does not hang around forever killing everybody; it diminishes in scope after a while. Why is that? Why don't plagues exterminate entire species?

The answers, as usual, are not simple—they include social and geographical reasons, among many others; but it's a force inherent in the infecting bug itself that I'd like to dwell on.

A bug that kills off every host shortly after infection is a very frightening bug indeed; from the bug's point of view, it's fatal, too. However nasty the disease it causes may be, a bug is not an evil entity. It is concerned not with wreaking havoc on people, but with propagating itself. The destruction of its environment (us) is, more often than not, an accident—not a beneficial one for the pathogen either, because a host that is dead or seriously ill does not come into much contact with other potential hosts and, thus, does not provide good infection opportunities for the bug. A dead host is commonly a dead end for its inhabiting pathogens, which will eventually die, too, not having anywhere they can escape to.

A successful pathogen is, therefore, not necessarily the most dangerous one, but the one that has found a way to spread itself continuously throughout the host population without undue fuss. This is why with most outbreaks, after a while, the infecting strain attenuates: It loses some of its more virulent properties and begins to produce less severe symptoms. Sometimes, the attenuated strain will, in the end, produce hardly any symptoms at all, and will manage to remain under the radar, with people passing it on to one another unknowingly. This is, of course, fine by the microbe. Thus, we find a microbe such as *poliovirus* (which causes polio), for which about nineteen of every twenty infection cases are subclinical (no symptoms appear). Before the advent of polio vaccines, it was very difficult indeed to be protected against the virus, because a large number of perfectly healthy people were unknowing carriers, too. If the virus had been more virulent, there would have been less opportunity for infection in the long run.

Attenuation is not the only force acting on the pathogen; our immune system also plays an important role. A healthy immune system can deal with a lot of trouble and can, frequently, destroy the invading pathogen. Once it has, the individual, having been exposed to the bug once before, is immune to its future attacks.[4] Even if the bug is too strong to be destroyed altogether, a healthy immune system can, frequently, contain and manage the troublesome invader, and keep it in check. In this way, not unlike the Cold War, a tense balance forms, giving both sides time to adjust to each other. Sometimes, one side ultimately prevails, and the pathogen is destroyed or causes rampant disease—this last outcome can occur especially if the host's health suffers from some unrelated problem, or the host just gets old and weak. But, sometimes, the situation just stays stable.

4. Which is, of course, how vaccines work: They expose the body to a harmless version of the pathogen, which enables us to build up a resistance to it.

All this leads to a scenario wherein a plague breaks out, kills people who are either immunologically weak (the ill, the elderly, or children) or simply unlucky, while people who are resistant survive and eventually come to form the majority of the population. Meanwhile, as time passes, the pathogen either calms down and becomes less virulent, or our immune systems learn how to handle it better (or both), and the plague eventually runs out of steam.

Why then, one might wonder, do plagues break out in the first place? If even the pathogen doesn't really benefit from it, why do they happen again and again?

The entire thing can, to some extent, be viewed as a biological misunderstanding between the host and the pathogen—or as a periodic escalation in the eternal arms race between the body's defenses and its attackers that occurs when a species of bug which had been under control randomly mutates and becomes too tough for the immune system to handle.[5]

The more problematic scenario, however, occurs when a strain comes into contact with a new, unprepared population—one that has not yet built up any immunological defences to it. This may happen when a virulent strain in one animal manages to jump to another species of animal (as avian flu did, for example); but, again, this is a rare event. A much more common event is two populations of the same species coming into contact, with one population having been engaged in an immunological battle with a certain pathogen for a long time, while the other population has never been exposed to it before. This is precisely what happened when Euro-

5. The flip side of this—a body finally managing to rid itself of a persistent low-key invader—also occurs but, because it doesn't manifest itself in any symptoms or in an outbreak of anything, we don't normally notice when it happens.

pean voyagers and settlers first came into contact with native people from other continents—Aboriginal people in Australia and Native Americans, for instance.

Simply by coming into contact with Europeans, non-European peoples suffered lethal, full-blown plagues of huge proportions. The pathogens that caused those plagues did not come from our evolutionary ancestors (those pathogens were naturally present in all human populations, because they were with us before any sections of humanity spread out of Africa), but from cows, chickens, pigs, goats, and the ever-pervasive rat—pathogens that only the livestock-raising peoples of Asia and Europe had been exposed and accustomed to. Because the infecting bugs had adapted to the Europeans' resistant immune systems, when they encountered the unaccustomed immune systems of the non-Europeans it was like running to force open a barricaded door that, upon impact, offers no resistance at all.

This explanation for the historic decimation of non-European populations by disease is, of course, a very sweeping and partial one. There are different types of immunity—innate and adaptive—and their different characteristics also have a bearing on how the pathogens affected those newly encountered populations. Then, there are the external conditions, such as malnutrition and stress (which the European invasions undoubtedly had a lot to do with), which would have strongly affected the vulnerability of their immune systems. Finally, other causes of death, such as war with the European invaders and famine, also took their toll.[6]

---

6. Alcoholism, another aspect of intercultural interfacing, bears much resemblance to infectious-disease dynamics, because populations that had not been exposed to alcohol consumption prior to contact with Europeans (and whose bodies did not need to evolve the physiological adaptations necessary to handle it) were struck very harshly by the effects of it.

At this point, we're entering difficult emotional and political territory. We know little of the actual immunological state of native inhabitants at the time that their lands were invaded, so biological explanations for these events are far from conclusive. It has been argued that to claim this "virgin soil" theory puts the horrors of invasion on a deterministic basis. In doing so, it absolves the settlers from guilt, because it implies that, no matter how good or enlightened they were, many native people would have died.

Another argument says that in looking at things this way, we accept, on an immunological basis, the idea of the supremacy of the white man. Others maintain that to hold this virginal view of pure "noble savage" indigenous peoples defiled by disease-infested Europeans is hopelessly romanticized and mounted on precious little evidence. Accusations of bias and misinterpretation abound, so I'll stop now, before I accidentally say something that gets me into hot water.

The Fight Not Won

Most of the time, we're not aware of the bugs that live in and on us. Once they make their presence felt, however, we fall ill and may even die. Throughout most of human history, resisting infectious disease was exceptionally difficult; in fact, disease was seen as a near inevitability. Medical science has been fighting infectious-disease-causing microbes for centuries—even though, for most of that time, they didn't know that microbes existed. In the last two centuries, there have been three great strides forward in our struggle:

1. Antibiotics (great stride).
2. Vaccines (greater stride).

3. Hand washing (greatest stride).[7]

I'm going to focus on antibiotics, because to do the story of vaccines justice requires at least another book, and I suspect you know the basic procedures of hand washing already.

The story of antibiotics started millions and millions of years ago. We humans didn't even exist back then. Microbes fighting each other for survival and resources developed chemicals for killing or repelling each other. This is still going on all over the place, of course, but over the past hundred years or so our involvement in this eternal struggle has taken a few turns.

About 200,000 years ago, humans evolved (originally as hunter-gatherers) and suffered from infectious diseases at some rate, as all living things do. Around 10,000 years ago, human civilization started gathering speed: We began to develop agriculture, and then towns and cities appeared, and this led to bigger populations and denser living areas. Because everybody was crowded together, it was easier to catch stuff from each other, and infectious-disease rates saw a sharp increase. New diseases also began to appear, most of them originating in domestic livestock (as we saw in the previous section) or in that ubiquitous urban pest, the rat.

Throughout history, bacterial and viral diseases were, hands down, the most common causes of death globally. Treatments (including rattlesnake pills, magical talismans, and lots of blood-based potions) were usually ineffective, and a person who caught a disease

7. You think I'm joking? Dr. Ignaz Semmelweis, possibly the most obscure hero in human history, was ridiculed when he suggested, in 1861, that doctors should wash their hands when moving from one patient to another. We learned, slowly, that he was right. Soap has saved more lives than any wonder drug ever could.

would have had to rely on his or her natural immunity and strength of constitution to survive.

Early Sightings

In the late-nineteenth century, improved living conditions and hygiene were responsible for lowering mortality rates in European countries. At the end of the century, sporadic reports emerged of molds that appeared to have antimicrobial qualities. The use of mold poultices for treating wounds had dated back thousands of years but, because the connection between microbes and human health was only established during that century, it was the first time that scientists understood, in principle, what the molds were doing. These reports were largely ignored by those who came across them.

Once humans had figured out that microbes could cause disease, there was a slow implementation of antiseptic practices by hospital staff, which brought down infection rates. Vaccines also began to come into use. Nevertheless, infectious diseases were still a huge problem for mankind.

At the beginning of the twentieth century, a few drugs that could treat infectious diseases began to appear: the arsenic compound salvarsan was used against syphilis, and sulfa drugs were being used to combat infections by the late 1930s.

What Is an Antibiotic?

An antibiotic is defined as any chemical substance that kills or suppresses the growth of microorganisms. However, to be of any medical use, such a substance also needs to be far less harmful to its recipients (humans or animals) than it is to its targets. This attribute—the ability to inflict damage selectively—is a very elusive one.

Because bacteria have a number of processes and structures that differ substantially from those of human cells, antibiotics can target them without doing damage to our cells. A classic example of such a structure is the bacterial cell wall, which has no counterpart in human cells. Its construction process can be inhibited by the antibiotic penicillin, so the bacteria can be targeted without affecting our cells.

A good counterexample also illustrates this attribute: Treating cancer is normally much more difficult than treating a bacterial infection, because the cells that we want to get rid of are rogue cells from our own bodies. The differences between the rogue and normal cells are therefore much smaller and more distinctive, and inflicting selective damage is much harder. Chemotherapy has to play on these small differences as much as possible—even then, it is really just a case of poisoning the tumor, hopefully, at a faster rate than the rest of the body.

For the same reason, antibiotics are no help against viruses or fungi: Fungi are fundamentally much more like us, and share more of our basic traits than bacteria do, so it is hard to target them without doing damage to ourselves. Viruses are a different problem altogether—although they are vastly different from any other form of life, they present a much more difficult target for drugs, because their means of reproduction is so intimately tied to the workings of our own cells. Until recently, the only way to deal with them was via the roundabout method of vaccination: that is, by presenting them to the body in a harmless form and counting on the immune system to take over when the time came. Currently, we have a small number of antiviral drugs in use, and fervent research (fueled by the need to combat the AIDS epidemic) is being conducted to find more.

Antibiotics are also no good in combating diseases such as malaria and leishmaniasis for the same reason: They are caused by parasites, which are, relatively speaking, our close relatives.

The Regulation of Antibiotics, Part 1

In 1937, an American drug company produced a raspberry-flavored sulfa drug. How nice of them. Less nice was the fact that it used the chemical diethylene glycol as a solvent. After more than a hundred people died of diethylene-glycol poisoning, it turned out that the company had not performed any safety tests to check its product. Worse still, the company had not broken any laws, other than by including the word *elixir* in the product name when, actually, it contained no alcohol. The latter act was illegal; poisoning people wasn't, and the company got off with a slap on the wrist. The outrage over this tragic farce led to the American Food and Drug Administration (FDA) being given a lot more power; since then, the FDA, whose standards are very strict, must approve every drug sold in the United States, which has significant global repercussions.

The Discovery and Production of Antibiotics

We now come to one of the most delicious events in science: In 1928, Dr. Alexander Fleming returns from a long holiday to his notoriously messy lab at St. Mary's Hospital in London. Among the many culture plates stacked in his disinfectant container, he notices one, which is fortuitously balanced at the top of a plate heap, and therefore untouched by the disinfectant. The plate has been contaminated by a fungus, and while the whole plate is full of an opaque mess of bacterial colonies, a circle around the contaminating fungus is clear of bacteria.

That single moment—when, amid all the things that were undoubtedly on his mind at the time, Fleming noticed one odd detail and understood its possible implications—is about as good as science gets.[8]

8. It has also provided hope and comfort to countless messy scientists, who've now gained a great excuse for leaving stuff around, instead of cleaning up after themselves.

Fleming isolated the fungus, correctly identified it as *Penicillium notatum* (*P. notatum*), and managed to isolate the antibacterial compound it produced, which he named penicillin. He published his findings a year later. They were mostly ignored by the few who read them.

This event brings up some of the most nagging questions in scientific and medical history: What place does luck have in scientific discovery? What if Fleming had been just a little tidier? What if that particular strain of fungus—a strain later found out to be a massive overproducer of penicillin, far beyond the usual rate for *P. notatum*—had not wafted into that particular lab? What if the weather had been different and had not encouraged the fungal growth? What if Fleming had returned earlier, before the fungus had time to make its visible mark in the plate? What if he had been preoccupied, or just hungover, and had not noticed the fungus? What would our world have looked like then? More practically, are things like that happening all the time, without our noticing, and can we improve our chances of spotting them?

Based on similar cases, and on common sense, it seems safe to say that penicillin would have been discovered at some point, and that medical and scientific developments would not have been radically different. What difference it would have made to history in general is anyone's guess.

It's worth pointing out two things: People had begun noticing the antibacterial properties of *Penicillium* fungi before Fleming (though he was the first to isolate the active ingredient), and Fleming himself, it appears, was not at the time (or in the decade or so that followed) fully aware of the larger potential of his discovery. He explored other antibacterial options that he judged to be more promising, made use of penicillin for modest diagnostic purposes in his laboratory and for treating boils, and did not pursue the clinical possibilities of penicillin or attempt too vigorously to interest the scientific community in it.

For more than a decade after its first purification, there was not enough penicillin in the world to cure a single patient. In 1939, researchers Howard Florey and Ernest Chain got curious and finally paid heed to Fleming's discovery. What followed was a large-scale effort to produce workable quantities of penicillin. Initially, production was just for scientific purposes; but when its full potential was realized, the U.S. Defense Department got involved. Large-scale production required trans-Atlantic collaboration (during wartime, when the Atlantic Ocean was a very dangerous thing to cross); another fair measure of good fortune, happenstance, and sheer concentrated hard work; and generous helpings of ingenuity from a large group of people.[9]

Fleming deserves much credit for his contribution, but the popular image of him as a lone savior of mankind does a great disservice to those working before and after him to bring the discovery to its applicable stages. A eureka moment, however brilliant, is never enough.

Near the end of World War Two, penicillin was widely available. It was initially provided to Allied soldiers for use (to replace the more primitive sulfa drugs, which had the added disadvantage of being a German innovation), and it saved many of their lives and cured many illnesses (including that prevalent combat-related condition, gonorrhea).

From then, there was a surge of research findings, and new antibiotics began to be discovered and manufactured regularly. The future looked brighter than ever. Humankind had finally achieved the

9. It also required a moldy cantaloupe. Out of 1,000 tested *Penicillium* strains, the one growing on a cantaloupe found in a fruit market was the best penicillin producer.

upper hand in the struggle against the natural world. The atom had been harnessed for power, outer space had come within reach, and mankind was now firmly on its way to ridding itself of the ancient curse of death and suffering from disease—it was a brave new world. In 1967, then U.S. surgeon general William H. Stewart is said to have stated that medical science could now "close the book on infectious disease" and turn its efforts to other afflictions, such as heart disease.[10]

In more recent times, the production of antibiotics has moved into the realm of the commercial world, where it responds to market considerations, such as ease of production, price, and stability. After all, a wondrous drug that expires in storage within a week and cannot be used, or that costs too much to make, is not much good to anyone.

The Empire Strikes Back

It turns out that the battle was not as simple as it seemed: The moon continues to move in its regular path and can be trusted not to wriggle away when we try to land on it for the second, or hundredth, time; an atom is a difficult thing to split, but atoms don't become any harder to split as time goes on; biological entities, however, behave differently—they adapt.

To our credit, we saw this problem coming: Researchers (including Fleming) observed, as early as 1946, that bacterial resistance to antibiotics was possible. To our discredit, we didn't do much about it: The newfound powers of antibiotics were so overwhelmingly useful that it was very difficult to be cautious. Besides, why bother? Bacteria

---

10. Just where, and to whom, he actually said this is unclear. Stewart said, in recent years, he could not recall having made the statement, and no one is able to find a good source. It may be one of those haunting misquotes—like Marie Antoinette's "Let them eat cake"—which ends up providing a good target for righteous indignation a generation later.

can't evolve *that* quickly, can they? Surely we could wipe them all out before they'd wise up.

Today, we know better: We know that there are major obstacles to eliminating an entire pathogenic bacterial species, and we know that bacteria not only evolve very quickly—when a generation lasts twenty minutes, there's a lot of room for evolving—they also help each other to do it.

We are also beginning to realize the full scale of the effects of HGT, which no one knew about fifty years ago: If one bacterial species manages to develop a gene that provides resistance to an antibiotic, other species can integrate that gene into themselves—they don't need to reinvent the wheel each time. Antibiotic-resistant pathogens, for example, appeared quite early on (especially in hospitals, where plenty of illness and plenty of antibiotics converge), and are now slowly but surely spreading out into the world.

It makes matters worse that antibiotics were (and, to a large extent, still are) often misused. Some doctors still give antibiotic medication inappropriately for illnesses that won't respond to them—usually to satisfy their patients' requests, to avoid being seen as uncaring, or as a blanket measure when the diagnosis is uncertain. Often, patients then stop taking their medication when they feel better, which is not necessarily when the harmful germs in their bodies have been exterminated.

The worst misuse of antibiotics, however, stems from pure economic interest: Antibiotics are routinely fed to livestock to prevent illness and to stimulate the animals' growth. This means that huge amounts of antibiotics (estimated at about 90 percent of all antibiotics administered globally) are given to perfectly healthy animals. This works well for the farmer in the short term, but it means that there are a lot of unnecessary antibiotics going around, which is a problem: If a germ is exposed to an antibiotic at a lower-than-lethal level, or for a shorter-than-lethal time, its chances of developing resistance to the antibiotic grow (especially if this carries on for several generations), in the same way that any adaptation to environmental conditions would occur. We're acclimatizing the germ to antibiotics, so to speak.

How Do You Solve a Problem Like Resistance?

Resistance to antibiotics occurs for many reasons: It can be because a bacterial enzyme is present that breaks the antibiotic apart; it can be because a modification in the antibiotic's target exists, so a drug that works by binding to a critical protein inside the microbe can no longer find that protein; and it can be because of a protein structure, called a multidrug pump, that can be found on the bacterial outer membrane. Also known as a molecular vacuum cleaner, multidrug pumps dump substances from bacterial cells—which means that after we have gone to the trouble of finding an appropriate poison to give the microbe, it inconsiderately chucks it right back out again, and goes on living. These pumps are very effective at what they do (not only in bacteria, but also in other types of cells, including cancer tumors) because they are not specific to any single type of substance, and they confer resistance to many types of antibiotics at once.

Countering antibiotic resistance isn't easy. There are two complementary approaches: The first is to let medical science play an offensive game, and the second is to let the rest of the world improve its defense.

Offense consists of finding better, newer drugs to work against other unique bacterial structures and mechanisms or against the resistance mechanisms themselves. Finding new drugs can be done either by testing various living organisms to find new active molecules (this is where antibiotics originally came from, remember?) or through rational drug design: creating computer models of the bacterial target so that we can design an artificial molecule that will latch on to it and inhibit its action. Rational drug design can be carried out in a lab and doesn't require trampling through the jungle and sampling thousands of molecules, but it requires good information about the target molecule at the outset, and some massive computing power. So far, it's been less successful than the traditional method, but give it time.

A simple, cunning treatment strategy using existing drugs that often works is the double punch: Give the patient two different antibiotics at once, in the hope that if there are germs in there that are resistant to antibiotic A, they will succumb to antibiotic B, and vice versa. This often works well, but not always: Multiresistant-germ sightings are on the rise, and our options run lower every time we try this.

The defense approach is more of a public issue: Increased awareness and more legislation to cut out unnecessary antibiotic use in humans and livestock is needed, and those measures lie outside the jurisdiction of the scientific and medical community.

The Regulation of Antibiotics, Part 2 (and Some Politics)

In the humans-versus-microbes game, the microbe team is quicker: They're playing as a team, and they have more experience. We're newcomers, and the only things we have going for us are our smarts, some sort of a game plan, and (potentially) a bigger budget for the season.

At the moment, though, drug companies are busy doing research on more profitable problems like heart disease, diabetes, cancer,

Alzheimer's, obesity, and impotence. Infectious diseases are still not enough of a problem to pour money into—put bluntly, not enough people are suffering for the market to become lucrative yet.[11] Government research has limited funds, and there's not yet enough public outcry for the politicians to pay much attention. If you remember me telling you before, the only way penicillin research got off the ground was when the U.S. government decided (after being persuaded by Dr. Howard Florey) that it was part of the war effort. A sad fact, methinks.

A closely related problem is the limited speed at which we are able to move when it comes to scientific development: The amount of time it takes to develop a new drug is measured in years, if not decades. There are always many promising leads to start with: I long ago lost count of the number of professional articles or research proposals I've read that end with the words, "These findings may have important therapeutic applications." The problem is that there are so many requirements to be met that only a tiny percentage of this innovative research ends up with a new drug or treatment on the pharmacy shelf or in the hospital ward, and the process of separating the useful from the useless is a long and winding one.

Microbes, on the other hand, are not only quicker to erect defenses; they're also unhindered by moral or political considerations. When a billion bacteria die and a single resistant one survives, it's par for the course for them. For us, even a single human death is too much. We have ways of protecting individuals—chief among them, government laws and regulations. Nowadays, before a drug is approved for use, it has to be demonstrated that it does no harm to anyone, which means

11. This is one of the limitations of the free-market system.

extended (and expensive) clinical trials. I wouldn't have it any other way (I don't appreciate being poisoned by corporations any more than you do), but it's a hindrance, nonetheless, and there's a delicate balance to be struck here between safety and urgency—something that regulatory agencies struggle with. Back in 1988, for instance, AIDS sufferers surrounded the FDA headquarters and shut it down for a day, in protest against what they felt was overly strict regulation that was delaying the approval of new antiviral drugs. We're dying *now*, they said, please let us use them; we won't be around later. The FDA agreed to fast-track drug approval in certain cases, but that tense balance of industry, community, and governmental interests remains.

Are We the World?

Meanwhile, in many parts of the world, the whole antibiotic revolution is yet to arrive, more than six decades after it began. Outside Western countries, people (especially children) are dying from the same old diseases—diseases that can be cured by existing, inexpensive medicines that are not made available to them for want of funding. Private benefactors and humanitarian organizations have recently decided to assume some responsibility for this situation by providing the means necessary for some of the research and treatment that is needed for these diseases. I think that's fantastic, and I hope it works. I also hope they're doing it wisely. Besides, seeing Bill Gates and Bob Geldof agree on anything is an unexpected pleasure.

Two Futures

What lies ahead? I tried looking up "prophets" in the telephone directory, but the closest listing I could find was "property consultants" (strangely fitting), so I'll have to give you the two alternative scenarios that I can foresee.

In the first, we see future generations suffering from resistant, widespread bacteria, again dying from diseases that we had all but forgotten, and blaming us for our selfishness and lack of foresight.

In the second, one or more of the avenues of research currently under investigation yield a better solution, and unborn generations are able to pity us for our primitive measures and for our suffering—just as we, today, pity generations past.

I'm rooting for the second future. It's quite possible that a solution that lies outside the box may help. We have made huge advances in immunology that have improved our understanding of how our bodies act against invaders, and we may find new ways to bolster these defenses. Other developments such as phage therapy, RNA therapy, probiotics (a buzzword used by yogurt manufacturers, but also a field with some potential), or something else entirely that we don't yet suspect could prove to be our salvation, at least temporarily.

The past has shown us that big scientific developments often come from humble, unexpected origins. Anything we come to learn about microbes may conjure the next big thing—who knows, the next wonder drug may even come sailing in through the window, as it did for Fleming.

Just What the Doctor Ordered

Research is all well and good for scientists; it certainly keeps them busy. What the rest of us can do, in the meantime, is simple enough. Aside from taking the correct medicine, like good little children, our best method of protection is the oldest one—we need to support the complex immune system that protects us every moment of our lives. The body is capable of defending itself rather well, and we need to give it the conditions to do so—simple ones, like eating well, exercising, and getting good sleep. It's not an insurance policy, but it's worthwhile.

Spiced Out

It is said that a wise man learns from the errors of others. Because I aim to make you, O reader, wise beyond your fellow human beings, I offer you these painfully acquired words of counsel: After ladling out green *zhoug,* do not lick the spoon.

Zhoug is a condiment that is Yemenite in origin. It's prepared from chili peppers, with garlic, parsley, and some other admirable stuff added in. It has one hell of a bite—especially the green variety. It goes well, if carefully used, with meat, fish, and other dishes. It does not go well with my digestive system when I mix an emergency-ward-admission-size gob of it into my hummus to make dinner more interesting; yet I do it time and again. Why dost thou tempt me, o devilish brew?

Adding spices, adding salt, preserving fruit in sugar and veggies in vinegar, smoking meat or fish, fermenting just about anything into alcohol, boiling, cooking, frying—what astonishing ways of messing with our foodstuffs we humans have devised over the ages.

You don't see other animals doing it: Squirrels have never been observed delicately peppering their nuts; cheetahs do not add cinnamon to their *antelope tartare.* To be sure, some wasps keep their prey (spiders) alive but paralyzed so that their young will have fresh meat when they emerge from their eggs; and the disconcerting tropical honeybee *Trigona hypogea* laces its food (semidigested meat from animal carcasses) with *Bacillus* bacteria to help its larvae digest it to perfection. They also appear to secrete antibiotic substances that keep this meat broth from spoiling in the tropical heat.[12] But otherwise, the most sophisticated culinary act I know of by a nonhuman

12. Antibeeotics, perhaps?

organism is to leave the food out, or buried, until it ripens a bit more. We humans, alone, have to complicate things.

The obvious reason that we season and prepare our food is because we can. It enhances the flavor, it eases digestion, and it makes food interesting. We have hands capable of operating a salt shaker, and we have domesticated fire—an innovation whose all-time best application must surely be the barbecue. But that reason doesn't tell the whole story. There is good reason to believe it's not a coincidence that all of these measures prevent microbes from surviving in our food.

This is a subtle point: We did not inherently like our food better when prepared this way; we have slowly, over thousands and thousands of years, been adapting our collective sense of taste.[13] We, as a species, have grown accustomed to the way that food tastes when preserved in these ways.

It's no coincidence, for example, that spicy food comes from hot countries. In places where the heat causes rapid spoilage, spicing things up is a good way to preserve food—not indefinitely, of course, but it helps for a short while, when there's no freezer to keep the food in.

Extreme conditions such as heat, salinity, or acidity are no shield against microbes; we saw them surviving in such conditions in earlier chapters. But it is illuminating to note that, as a rule, microbes that can survive extreme conditions are not disease causing. A microbe causing an infectious disease is, almost by definition, one whose optimised growth conditions are very near the conditions found inside the human body. Makes sense, doesn't it?[14]

---

13. There are, of course, regional and personal variations.

14. A well-known exception is sporulating microbes: They can survive in harsh conditions inside their spores, then revert to their active forms upon infecting the host.

You may have been wondering why I included sugar in the preservative list. Since when is sugar an antimicrobial substance? Am I on some dodgy lollipop-manufacturer's payroll?

The thing to understand here is that different concentrations of substances affect microbes in different ways. Microbes prefer certain conditions: The barrier between their inner environment and the rest of the world is very thin, and their connection with the outside world is therefore very intensive. An environment jam-packed with sugar or salt wreaks havoc on their metabolism—water will literally be sucked out of them by the high concentration of solutes outside. In reduced concentrations, sugar and salt cease to become a threat, and are highly sought after by microbes. Tooth decay, for example, is brought on by microbes dwelling on your teeth while they wait to eat the sugar that lies there; they then hang about excreting acid, which dissolves the protective enamel on our teeth—it's just a side effect of their lives; they don't mean any harm (a fact that will surely provide you much solace, come your next dental bill).

Spices, on the other hand, have a different relationship with microbes. They are found in the living plant, where they defend it against invading microbes in a variety of ways. What we are doing when we spread basil or oregano onto a dish is lacing it with natural antibiotics. Garlic and onion are at the top of the list when it comes to antimicrobial strength—and consuming them also prevents disease in another way: They keep people from coming too close, which minimizes your chances of catching any flu, cold, or STD that's passing by.

BONUS TRACK #5

My Beloved

How could I have done this to you? We had been together for so long. Twelve years have passed since we first met. I do not remember where it was that we first chanced upon each other, but you have been with me ever since.

You know, as well as I, that ours was a troubled union from the start. Even in those early, heated days, I often wished you would go away. I hid your existence from all who knew me—a tryst like ours is frowned upon by all; I could not bear the thought of exposing it to the public eye, and our time spent together was always kept intensely private.

Oh, but what joy those moments provided; the pleasure of the flesh made doubly sweet for being illicit. In what ecstasy were those brief stolen moments spent; and yet, I would repent after each rendezvous. "Never again," I would rebuke myself, "shall I give in to these base, animal cravings. It is a doomed affair that will bring me nothing but trouble."

At times, I had convinced myself that you were gone from my life for good—a thing of the past, to be recalled with a soothing mixture of nostalgia and repentance, when remembered at all. But time and again, you returned to me, beckoning with the summer breeze, and I acquiesced. I can say, in my defense, that I never once strove to hasten your arrival or wilfully prolong your stay; but I know full well these are paltry excuses.

At first, you were modest and unassuming—content with your place in my life; but as the years flew by, you grew more demanding, your visits more heated, and the marks they left more indelible. I tried to reason with you. I said, "I am now a

continues

taken man—I cannot do this any more." Yet you persisted, and I was all but helpless against your tingling siren song, when it would come.

The final straw, my dear, came when you turned up again, last summer, fiery as ever. That season I will hold forever dear, for it was a passionate time, and my pleasure reached heights that I had never before known with you. Still, you grew ever more possessive. You would not stay where you have always been, you wanted to expand, explore. You needed to claim more of me for yourself than ever before.

I would not have it. The affair had run its course. I was no longer the carefree youth that I had been when we first met. I tried to let you down gently; to make a clean, amicable break of it—but you would not have it. I do not blame you, for I should have been stricter at the start, never should have led you on. I am sorry, my dear. You left me no way out but the final one.

At last, I did the deed. I turned my heart to stone, turned a deaf ear to your pleas, and smothered you repeatedly, until you were no more.

You are gone now, *Tinea pedis*, gone from my left foot forever. The antifungal cream has done its job well; yet I remember you with fondness. Involvement with the likes of you was all but taboo in our society, and any hint of enjoyment gleaned from the experience unmentionable. You are a thing to get rid of, not to savor—though I suspect that I am far from alone in my sin. I sometimes catch myself contemplating that spot you held, there between my second and third toes, and the guilty, heady tang of the forbidden once again echoes in my memory. I shall not forget you.

CHAPTER 6

Bugs on the Job

When I imagined a world without microbes at the beginning of Chapter 5, I mentioned something about the total obliteration of life as we know it, before quickly moving on to the disposal of white goods and toothbrushes. I may have been a tad too hasty in dismissing this subject. It appears that life, the odd stubbed toe notwithstanding, is a good thing to have. In fact, research shows that people who are alive tend to be more creative, perform better at their jobs, and make better romantic partners than their nonliving counterparts.

Life was developed, and is continuously maintained, by a large team of ceaseless, single-celled support staff. It is a simple point, but one that occasionally needs to be hammered in: Life on Earth relies completely and utterly on the existence and actions of microbes.

All That Matters

There are over a hundred different elements in the universe, but life on Earth is mainly based on four: carbon, hydrogen, nitrogen, and oxygen.

All of life—all things chirpy, slimy, fuzzy, buzzy; all things flying, swimming, gurgling, infecting, procreating, litigating—are made up mostly of combinations of just four kinds of atoms. We humans categorize these combinations based on their structure and their function, and we give them names: proteins, DNA, RNA, carbohydrates (sugars), lipids (oils and fats)—the building blocks of life. They're all, when we get down to the nuts and bolts of it, different varieties of carbon chains, interspersed with and surrounded by atoms from the other three elements, plus the occasional atom from one of a few dozen other elements (which include iron, manganese, calcium, magnesium, and potassium, to name a few) that are strategically placed at certain locations for maximum effect. These other elements are very important, but are only needed in trace amounts; for example, a single iron atom forms the core of every hemoglobin molecule in our blood. Without it, hemoglobin can't bind oxygen, and we suffocate.

All living organisms also need energy, which they obtain either by eating other living things or directly from the sun. This energy is used up quickly—much of it powering motion (in humans, this means muscles) and other bodily functions. In any process, there is never 100 percent efficiency (though it should be said that precious few human-designed motors aspire to the efficiency of our physiological processes), and a large part of the energy is lost, mostly via heat.

Humans and animals mainly get the substances needed for building and maintaining our bodies (to replace everything from the dead skin cells we shed to the carbon that we emit when we exhale, in the form of carbon dioxide) from our food.

By contrast, planet Earth is pretty much a closed system: It receives nothing from the outside but sunlight and the occasional paltry meteor. Living creatures are intertwined in endless cycles that see different molecules and atoms constantly change shape, form, place, and role. It is the dance of life, and its beauty is wondrous to behold, once you understand what you're looking at.

This cycle of life is why I harbor some extra-strength resentment about the term *food chain.* If something is consumed in a chain, a series of one-way links, it eventually runs out. Life on Earth is more like a very elaborate battery-powered gadget: The sun is the battery, so to speak, and everything else moves about inside it in complicated circles. This is not a chain—it's a full circle in which the topmost predator is eaten, in the end, by smaller creatures.

The role that microbes play in all this is critical. To begin with, they set up these cycles, back when they were the only game in town. Bacteria were the original solar panels: They carried out the astounding feat of capturing solar energy inside molecules (and, as we've seen, releasing oxygen into the atmosphere as a side effect), providing energy that the rest of Earth's inhabitants could, in turn, use. Nowadays plants (their multicellular descendants) do this, too, but photosynthetic bacteria are still responsible for trapping most of the sun's useable bounty. Then, at the other end of the cycle, different microbes (along with a host of creepy-crawly things) return the elements and compounds that the rest of us need back into circulation.

The life cycle doesn't only involve living things: A large bulk of its vital elements are not inside living organisms at any point—for example, most water is in oceans, lakes, and clouds, rather than inside bodies; most iron is in the ground; and most nitrogen is in the atmosphere. These elements form our biogeochemical cycles (cycles that draw from geological and biological bodies); but in order to bring them into the cycle of life, a fair number of critical chemical reactions are necessary that only microbes (for their own very good metabolic reasons) perform. The rest of us rely on these microbes for a steady supply of useable molecules to keep us going. Sulfur is a good example.

What a Gas

Bacteria use sulfur compounds in so many different ways that the mind boggles. Take, for example, hydrogen sulfide (H_2S), the stinky gas that we wrinkle our noses at whenever someone farts. Some bacteria can use it for oxidizing organic molecules (what this means, in plain language, is that when no oxygen is available for breathing, these bacteria use sulfur instead). Others use it instead of water, during the process of photosynthesis (H_2O versus H_2S—can you spot the similarities?). Still other bacteria use it not as air or water, but as fuel.[1]

In bacterial bodies, as well as in our own, a wee bit of sulfur is indispensable for maintaining the structure of just about every protein molecule. Plants and microbes take sulfur from the soil or the atmosphere, and incorporate it (or "fix" it) into amino acids, which are the building blocks of proteins; something then comes along and eats the plant or microbe; then something eats the first something that ate the plant or microbe; and so it continues, with the amino acids "traveling" from organism to organism until they reach, say, me. When my time comes and my body is laid to rest, my sulfur-containing amino acids will be dismantled back into H_2S by bacteria. On a less somber note, the breaking down of protein into H_2S is also done in our intestines, by bacteria. Blame them for that suspicious smell wafting through the room.

---

1. Any attempt to explain this further will rapidly escalate to the deployment of the term *oxidation/reduction potential,* which, as the slash clearly signifies, is not for the faint of heart. Explained bluntly: All the chemistry of life ultimately boils down to the direction in which the electrons are headed.

A New, Improved Sheep Burp

It is a well-known fact that, unlike humans, cows (along with sheep and other ruminants) can digest grass. It is a less well-known fact that cows (along with sheep and other ruminants) *can't* really digest grass.

The only reason for bovine superiority in the field (sorry) of grass digestion is that they cheat: They have, hidden away inside their rumen, an elaborate team of specialist grass-digesting microbes that expertly disassemble cellulose—the tough, hardy fiber that makes up plant-cell walls and that makes plants indigestible to humans.

Imagine if we humans had that—we could have done away with the middlecow and gone straight to the source. As it is, we'll have to stick to the steak and keep off the grass.[2]

Living among these cellulose decomposers in the ruminant gut are methanogens—a class of archaea that produce methane gas as a by-product of their activity. The methane is not used by anything else inside the ruminant, and is expelled from either its front or its rear end. This is a problem: Methane is a very potent greenhouse gas, and methanogenesis is a major environmental issue. (For example, sheep burps account for about a half of New Zealand's annual greenhouse-gas emissions.)

Scientists are working on solutions. One research group is developing a vaccine to drive the methanogenic microbes out of the rumen. Another solution may come bouncing in from an unexpected direction: Kangaroos are ruminants, but they don't produce methane. Instead of methanogens, they have gut microbes that produce acetate, which isn't a greenhouse gas at all, and can be used by the animal as an energy source. If researchers manage to replace the methanogenic archaea in livestock with these 'roo 'robes (not an easy

2. Or disregard cows altogether and become a vegetarian.

task, mind you), they'll have a more environmentally friendly sheep (or cow), with an added energy source, which will save the farmer some feeding expenses. Plus, they'll have achieved what must surely be the most Aussie solution ever.

Coal-Powered Laptops

I like the R.E.M. song "Shiny, Happy People." It's a good song, and it comes with an excellent video clip that shows a group of people dancing and hopping about excitedly, while a screen with shiny, happy pictures rolls behind them, powered by an old man laboriously pedaling a sort of stationary bicycle.

The image has stayed with me. When I look at something, I try to peek behind the foreground gloss and think of the workings behind it. Sometimes, this is not such a good idea—absolutely no good will come from imagining what the kitchen looks like while you dine at a restaurant. As a general rule, though, I recommend a frame of mind that keeps things real—it serves as an antidote against the fanciful notion that we can get something for nothing, which tempts us at every turn.

Think of the most advanced gadget you own. Maybe a laptop? An MP3 player? A GPS device? Now say to yourself, "This thing runs on coal."

Unless you've switched to a green-energy power provider or are receiving your power from your friendly neighborhood nuclear-power plant, the source of the electricity powering your home and recharging your mobile phone is most likely coal. A coal-powered computer—how absurd it is to picture such a sleek, modern thing connected to a fume-belching furnace; yet, save for a few intermediates, that's exactly what is happening. The only reason we don't see this is because we have wisely moved our furnaces far away, beyond the edge of town, where we have engineered turbines and generators to convert the coal to electricity, and strung out wires to deliver the power back to us. Without coal, and without somebody to dig that coal out of the ground and deliver it to us, the whole system would shut down.

I have no intention of getting into the humanitarian aspects of all this; wiser people than I have been in extensive discussions over the politics of the resources and labor that this demands. I do, however, wish to peek a bit closer at the source itself.

Coal is a form of long-dead organic matter which—apart from the occasional dinosaur—is composed mainly of microbially degraded plant matter that has been buried in the Earth and subjected to high pressure and heat over hundreds of millions of years. Petroleum and natural gas are also the remains of (largely microbial) dead organic matter—your car burns bugs for fuel! Because plastics are derived from petroleum, we can add them to the equation, too. Our economies, technologies, and lifestyles are based on consuming these huge, ancient deposits of plant and microbial remains by converting them to energy and using them for our needs.

I'm not suggesting that we should feel bad about it or anything. There's also no need to feel any gratitude or appreciation toward our demised benefactors (they're plants and microbes—and dead ones at that). I only mention this so that we can understand and remember what the foundations of our current lives are and, also, accept that these foundations will, at some point in the future, begin to run out. The rate at which new organic matter is accumulating is minuscule compared with the rate at which we're using it; we should be thinking seriously about starting to replace our energy sources gradually—so that our shiny, happy lives, with their dead-microbe-powered comforts, will not grind to a halt.

The Mess We're In

Close your eyes and try to imagine a world that does not contain plastic bags.[3] When you're done, have a go at imagining life without plastics altogether. Now open your eyes and try to count each and every plastic item, or part, that you see around you at this moment, wherever you are.

During the last fifty years or so, human technology has brought forth into the world a totally new class of materials. Because they're so new, and because they've been designed artificially, they're not a part of the regular patterns of nature. On one hand, that's a good thing—we don't want fungi eating away at our things—but they pose a new problem: How *do* we break down the enormous amounts of plastic waste that we continue to generate?

The problem gets even worse when we turn to artificial chemicals that are active, rather than inert. A plastic doll or saucer may not be degradable, but it's made of inert chemicals that do no harm to the body; however, there are things made from artificial molecules that are active, and these will damage us, and other forms of life, either as a side effect of their function (paints, solvents, and industrial waste, for example) or as part of their purpose. Pesticides are a prime example of this last category: An efficient pesticide needs to be harmful to the pests, but harmless to the plant on which it is sprayed, to the person spraying and picking the crop, and to the person eating the produce. That's rather a tall order, and it's not surprising that early pesticides, such as DDT, did not live up to the task, thereby creating health and environmental problems when used extensively.

---

3. Ignore this instruction if you're reading this book while driving.

What's worse, pesticides, along with chemical fertilizers, stay in the soil or seep into groundwater. Add industrial waste to this unholy soup, and you get trouble that just isn't going to go away.

This situation isn't black and white. Abolishing pesticide treatments and the use of chemical fertilizers would seriously set back agriculture: Farmers would suffer huge losses, and the price of food would go up dramatically. In poor countries, it might mean the difference between nutrition and malnutrition for a considerable number of people.[4] What the financial and, more importantly, the human price will be if everyone went organic is beyond me, but it's going to be dear.

Is science to blame for this mess?

On the face of it, it is. Scientists and engineers were the ones who developed these harmful substances in the first place. But we need to remember that science is a method of thinking, not an organization. It has no governing body and no central directives. Blaming science is meaningless. Blaming technology is even worse—it's a mere word that applies to everything from fire to nanotubes, to a guy pottering about in his shed with a wrench. Assigning responsibility to science, as some people tend to do, is just a way of avoiding having to think about the deeper issues; it goes no distance toward solving current problems and avoiding future ones.

We should learn from the pesticide conundrum (and others like it) by observing where the scientists went wrong: where they did not have enough foresight; where they were not restricted by the necessary ethical guidelines; where their judgment was clouded by other

4. The situation is similar to the "antibiotics in animal feed" bit we talked about earlier; however, because people can manage to live without meat, but not without plant food, it's a lot more critical.

considerations; where better government regulation was needed to override the financial incentives of research companies and the farming industry; and where public awareness should have been higher, and how it should have been raised.

Once we're done with all that breast-beating and self-improvement, we might also try to rectify the present situation, which is exactly what the research field of bioremediation seeks to do: It looks for ways to convert harmful compounds that are not naturally degradable into innocuous products. We can also do this using chemical processes—but microbes, which are alive, self-reproduce and, having a large array of enzymes at their collective disposal, usually do a better, cleaner, and cheaper job of it.

Considerable efforts are being made all over the world to find, isolate, and study strains of bacteria, algae, molds, and fungi that have the capabilities we need to biodegrade our toxic waste. Incredibly, for nearly every chemical that human ingenuity has devised, there exist organisms able to deal with it. A fungal strain eats nylon; a bacterium feeds on polyester; even the resilient polystyrene may be handled by the super-degrader *Pseudomonas putida*. All but the most unmanageable (or "recalcitrant") synthetic materials can be broken down by some enzyme in some microbe—the trick is to find it.

A good place to start searching for the right microbe is within the waste itself; the process of natural selection makes it probable that a microbe thriving in a toxic environment has found a way to tolerate it—that which doesn't kill them makes them stronger, it appears.

But isolating and locating the right microbe in a petri dish in the lab is only the beginning of the solution. The next step is to scale up the process by creating an industrial-scale bioremediation plant that is cost-effective and financially backed, and to convince everyone to sort their waste out and deliver it there for processing. Each step is fraught with difficulty, even those that don't involve people: Just because a fungus digests a certain type of plastic doesn't mean that it does it efficiently enough, so we have to find strains that digest even faster (like the strain of *Penicillium* found growing on a cantaloupe), and then nudge the existing strains into evolving better degradation capabilities—or even extricate the relevant genes and insert them into another type of microbe (genetic engineering at its most tree hugging).

Mixed waste is another huge problem. In Chapter 2, we saw the way that a microbe (*D. radiodurans*) can deal with mixed-chemical or radioactive waste; but even garbage as commonplace as the contents of an office trash can needs several separate treatments for complete degradation (it's very hard to get laser-printer ink out of paper, for instance), and the ideal conditions for one type of degrading microbe may be unbearable to another type. The next time you argue with coworkers over the office air-conditioning, or with family members over the volume of the car radio, be thankful that you, at least, belong to the same species.

Bioremediation scientists also work to ensure that biodegrading microbes don't break out and eat our things. They also concern themselves with water and soil purification, heavy-metal contamination, radioactive waste (microbes can't quench the radiation, but they can help prevent contamination, which helps a bit), oil spills, air quality, and much else that ails our world.

Ultimately, though, bio-measures alone, however successful they may be, will not ensure humanity's future. An invention that reduces the environmental cost of driving a car is useless if we all use it as an excuse to drive more. Scientific solutions to global problems must always be applied with wisdom, and must come with adjustments to thinking and behavior within society. In short, it would be a bad idea to count on technology to clean up our mess.

The Ultimate Working Class

Biotechnology is a catchy buzzword. I thought I knew what it meant, but I was surprised when, at the start of my first biotechnology course, the professor defined it as "the industrial use of microorganisms for the manufacture of a desired product."

His unspoken message was clear: Biotechnology is not merely a convenient description to be used when we want people to be impressed, or to invest money in a project that has some relation to biology or medicine (and that probably uses computers). But the professor was fighting a losing battle. Biotechnology (especially in its sexier form, *biotech*) has become a popular catch-all term, nowadays, for all applicable areas of the life sciences.

Quibbling over definitions aside, what are some of those industrial uses of microorganisms for the manufacture of a desired product? There is a lot of work being done for us by tireless microbial workers. The advantage that biotechnology has over other means of production is that it harnesses the very force of life for our benefit. The principle (as with bioremediation) is to make use of the microbe's natural drive toward survival and reproduction, so that we only need to encourage it in a direction we'll find useful. If you'll allow me to wax metaphorical for a moment, it is something like a canoeist going downriver: She requires the skill to control her craft accurately; the strong current will provide the power.

The use of microbes in industry is not new; we've been fermenting wine and beer, baking bread, and making yogurt (all microbially based processes) for thousands of years—long before we knew of the existence of microbes. The modern age of biotechnology began during World War One, when Britain desperately needed large quantities of acetone in order to produce cordite—a propellant for ammunition in artillery shells. An exiled Russian chemist working from a converted basement in Manchester made the first industrial biotechnological plant possible, using bacteria to produce acetone for His Majesty's armed forces.[5]

By the 1940s, penicillin was being produced in mass quantities, and scientists were getting excited—biotechnology's potential for the efficient generation of products had given rise to hopes that some of society's problems, from world hunger to energy crises, could be solved. That didn't turn out to be the case, but a large number of industrially important products (such as citric acid and malate, which you consume regularly, but may not have heard of) were, and still are, made by microbes.

In the 1970s, new techniques were developed that enabled biologists to move a piece of DNA from one organism to another. The technique, called recombinant DNA technology, took biotechnology to a whole new level, and led to a parallel leap in safety concerns and public worry (some of it justified, some less so).

A dramatic success came in 1978, when the gene for human insulin was inserted into *E. coli*. Prior to that, people suffering from diabetes would receive pig or cow insulin, purified from the animal's pancreas. This was expensive to produce, and sometimes caused allergic reactions. *E. coli*–produced insulin is cheap and nonallergenic, and I'm guessing that the pigs are also a bit happier now.

Nowadays, biotechnology allows us to produce all sorts of industrial and medical compounds using microbes, including vaccines, hormones, antibiotics—the list goes on.

---

5. Via a complicated sequence of happenings, this event also led to some interesting nonscientific developments, including the establishment of Israel a few decades later with the very same chemist, Chaim Weizmann, as its first president.

Special Delivery

Cancer, you don't need me to tell you, is a nightmare. Because it is, in essence, a betrayal—one of our own cells renouncing its proper place in the grand scheme of things and multiplying without restraint—medicine finds it very difficult to cure: The differences between the renegade cells and our normal cells are so small. Chemotherapy and radiotherapy are our best treatment options, but they still cause grievous harm to the body. We need something that can target cancer cells specifically. That's where we may find help from an unexpected source—our ancient enemy, the virus.

Viruses aren't generalists. A virus infecting a human body is able to attach, enter, and replicate only in specific types of cells. A flu virus, for instance, will infect respiratory cells of a certain type, and will not be found in liver or brain cells. Could we use that specificity for therapeutic purposes then? Create a kind of "magic bullet" solution to one of humanity's worst afflictions? It's a compelling idea—take a virus, infect the person, and get rid of the tumor.

It's not that simple, of course. First, we must either find or engineer a virus able to infect the tumor cells; then we need to ensure that it will infect only (or at least mostly) the tumor cells, or we'll have a viral infection on our hands. The virus also needs to slip undetected through the body, so the patient's immune system doesn't destroy it. Cancer tends to mutate rapidly, and anticancerous viral therapy will need to address that, too. There are clearly many obstacles; however, there are also glimmers of hope. One British research group, for instance, has started clinical trials with a "stealth" *adenovirus,* which is essentially a modified cold virus coated with an artificial polymer that allows it to travel safely through the blood to its destination. Once a single virus particle infects one tumor cell, it sheds its coat and replicates inside it, bringing forth millions of specific killers to the immediate vicinity of tumor cells. That particular avenue of research should be interesting to watch.

Viral messengers can also be used to deliver drugs to the exact place they're needed, which eliminates the need to drench the body with a drug, when only a small percentage of it will end up where it should (which is what we're doing now every time we take our medicine). This shouldn't even require an entire virus: We can take just the few proteins that determine the virus's specificity, tack them onto a shell of our own devising, and send it away. This reduces to virtually zero any chance of a virus doing unwanted damage.

The most daring application of viruses, however, occurs in targeted gene therapy: For people with innate, inherited genetic disorders (a faulty gene), a viral-based system may be used to insert a working copy of that gene into the specific cells that need it.[6] Perhaps the day will come when conditions like cystic fibrosis or color blindness could be cured by a deliberate therapeutic virus infection.

Mini Miner

Would you like to know how to make gold from water? I'm not selling that knowledge cheaply; first, you'll have to hear about other metalworking microbes—we'll start with *Acidithiobacillus ferrooxidans* (*A. ferrooxidans*).

A. ferrooxidans burns metal and feeds on air. This ubiquitous bacterium can fix carbon dioxide and nitrogen from the air for its own use. That in itself isn't so special: I introduced you to other nitrogen

6. Recent breakthroughs in stem-cell research also use viruses to insert reprogramming genes into cells.

fixers in Chapter 4, and assimilating carbon dioxide is a trick that every photosynthetic plant or microbe can manage, given sunlight (*A. ferrooxidans* isn't photosynthetic, by the way).

What makes it interesting—not only to scientists but also to businesspeople—is that it can oxidize metal to get energy, in the same way that our bodies oxidize glucose. Given the right conditions, when this bacterium is sprayed on metal ore, it can extricate the valuable metal from the ore—a process called bioleaching. The conventional method for extracting metal from ore is chemical, not biological, but humans have been mining metal for a while now, so the best deposits are long used up, and ore quality is steadily getting lower—which is where bioleaching becomes more commercially viable.

The method has its problems—microbes are slower workers than chemicals, for one—but as the know-how gets better and ore quality goes down, over time bioleaching's popularity is going to grow. Bioleaching processes are already used extensively in copper mining, as well as in the mining of other valuable metals, including zinc, nickel, and uranium.

Bacteria have also been useful in the mining of gold. Gold is occasionally found stuck firmly (occluded, to use the technical term) inside certain minerals. A number of different microbes can perform a similar oxidation on these minerals, converting them into soluble ions, and thereby exposing the gold inside. It's not as if there's a wee nugget hidden inside there, though—this exposure is still at the level of single atoms; but now, the standard chemical processes can be applied for extracting the precious metal. The microbes may not swing the hammers or operate the heavy machinery, but they can save a lot of chemical tinkering, and keep low-grade metal ore commercially exploitable.

It's a hard way to get rich, all that digging and extracting. Thankfully, gold is also found in more easily reached places, like in ordinary water. Yes, there's a tiny amount of gold dissolved in there—all we have to do is find a way to concentrate it, and we'll be rich! *Ha ha!*

One lucky bacteria, *Pseudomonas stutzeri*, knows the secret. This bacterium is able to oxidize the gold ions floating around, and convert them into solid gold—a living philosopher's stone, even better than the one that medieval alchemists searched for in vain. Sadly, this application is not very profitable. The amounts of gold that can be recovered in this way are far too small to bother with. We'll just have to find some other way of making those millions.

Bzzzt

I used the word *oxidation* a few times in the last section. It's a technical term that means "transfer of electrons." Getting rid of electrons is something that microbes, like all living things, need to do all the time.[7]

As the word suggests, oxygen is a good acceptor of electrons and is frequently involved in these reactions. However, sometimes oxygen is unavailable, and microbes find other ways to oxidate, such as alcohol production, among many others. Quite recently, another way was proposed: Recent findings suggest that certain bacteria living deep inside the soil, where there is no oxygen to speak of, grow long, very thin fibers called nanowires, whose role it is to move surplus electrons, and dump them into metal deposits.

This may sound oddly familiar. Moving electrons are also known as electric currents, and the fibers that transport them are commonly called "electric wires." Because this finding was only possible because of newly acquired microscopic techniques, there's no telling how common these things are. It may be that the soil under our feet is a

7. You may have done so yourself, if you've ever breathed.

veritable buzz of electrical activity, while further speculation has gone as far as suggesting the possibility of communal microbial power grids.[8]

Researchers are thinking that perhaps we can use these nanowires for making very small electric circuits, or even for making microbially produced chips. It will take some time, though, before we even find out if growing a computer chip is possible, let alone actually doing it.

At a less speculative stage is something called a microbial fuel cell. This is a device that works on the principle that we can use the microbial capability for moving electrons to power up things of our own. If we stick one electrode in an environment where microbes desperately need electrons, and connect it to another electrode in a place where there is a surplus of electrons to be found, we will have a battery that will last as long as those microbes live there—which could be forever.

These fuel cells are still a very weak source of electric power; but, even in their present form, they may be useful for powering things like deep-sea underwater monitors (changing batteries at the bottom of the ocean is a hard job, after all). In the future, if researchers and engineers can improve the technology, this could be huge—energy could be produced from sewage, food scraps, poisonous gases, or just about anything. Power from trash and pollution? Good one.

Another use for microbial fuel cells would be to replace the batteries that power devices implanted in our bodies—tiny electronic devices that can stimulate nerve cells to help restore hearing and vision, or help sufferers of Parkinson's disease and spinal cord injuries. A Canadian team has recently announced that it's developed a pro-

---

8. Because no one reads footnotes anyway, I'll indulge myself with a truly outrageous science-fiction notion: Electric networks are used not only for power, but also for communications. Our brain is a collection of cells interconnected by conductive fibers. Think about it.

totype of a fuel cell that uses a strain of yeast (*S. cerevisiae,* the common baker's yeast) to produce electricity—it feeds off a bit of the glucose that runs naturally in our bloodstream to generate small electric currents that could keep these devices running forever. This is still at the "proof of concept" stage (techspeak for "don't wait up") and plenty of problems are yet to be solved, but if it works, friendly electric vampire yeast could eliminate the need for costly and inconvenient battery replacement surgery.

Our last glimpse into the future comes from Japan where, in honor of Einstein, the message "$E=mc^2$ 1905!" was translated into binary code, "written" in DNA base-pairs, and inserted into a bacterium in order to show that data can be stored in bacteria. Living memory modules may have some interesting advantages: They cannot be erased like the magnetic media we currently use, and they copy themselves. If error-prevention safeguards and other complications can be ironed out, we may someday be saving our files in strange, wriggly formats.

Moreover, the technique could be used in the future to write into a genetically modified organism, in plain language, what the thing is—something between using annotated computer code and tagging a bird. Perhaps a future scientist or doctor may isolate a bug, sequence its DNA, and read, "This strain was developed by Prof. I. M. Smart, SmarterLabs Ltd, Oct. 2032." Or, if things get out of hand: "Special offer—50 percent off all SmarterLabs products for holders of this microvoucher." Maybe we'll even be sending love bugs to each other—*Say it with DNA!*—or, for the ironically inclined, a get-well-soon message encoded into the bug that causes the illness. It's the thought that counts.

BONUS TRACK #6

BioFuture

Biological data storage is a nice concept but, apparently, it's not revolutionary enough for some: One of the hottest current buzz terms is *synthetic biology*, an avenue of research that strives to create and reengineer life from its simplest building blocks.

Traditionally, it was mostly a way of understanding how existing life works but, as DNA synthesis and sequencing techniques become faster and cheaper, engineers have begun to think along more radical lines. Instead of using naturally occurring genes, these engineers are talking about constructing small, standard, functional modules of biological information, and piecing them together into networks, like electronic components of a computer chip. Engineers are not ones for extended contemplation—they've already rolled up their sleeves and begun producing "biobricks," which they've assembled together to produce the world's first biological "software."

The "hardware" that these DNA programs run on are the simplest forms of life—bacteria, naturally. An engineered *E. coli* that flashes on and off under ultraviolet light to demonstrate a simple repeating binary switch action was constructed back in 2001. Since then, some whimsical and strange creations have followed, many still at the proof-of-concept stage, or only slightly beyond. My favorite is an *E. coli* that has been engineered to exude the scent of minty bananas, instead of its usual, um, piquant odor—an invention that might make biological labs the world over become an even more ecstatically pleasurable place to work in.[9]

---

9. If we could somehow introduce these sweet-scent genes into our natural gut flora, imagine the consequences this would have on the experience of going to the bathroom or—something a certain twenty-pound member of my household keeps me aware of—changing diapers.

The possibilities that this new discipline hints at are boundless, as is usually the case with things that have not been tried out yet. Rewriting the genetic code, churning out new materials, and nanofabrication to order—exciting stuff all around. Curbing the enthusiasm are the usual worries about biosafety, bioterrorism, biowarfare, bioethics, and general bio-messing-with-creation—worries that the synthetic-biology community continues to make reassuring noises about (though I'm sure they wouldn't object to some external directives, considering their expertise with regulatory systems). The engineers behind these creations also point out that, unlike regular genetic engineering, synthetic biology will be able to create and use systems that work very differently from naturally occurring ones and should, therefore, pose no threat to humans, making it even safer to pursue them. Hmm.

What practical result all this may lead to is anyone's guess. Synthetic-biology visionaries conjure images of self-repairing bodies; photosynthetic humans; houses that grow out of seeds; and specially developed microbes that clean up the atmosphere from pollution, and colonize Mars and Venus, in preparation for human settlement. It's easy to scoff at such pie-in-the-sky notions but, sitting here in my room, calmly tapping on a two-pound machine that has no wires attached to it whatsoever—a machine made of materials that did not exist a few decades ago—chatting nonchalantly with a friend halfway across the globe using signals that fly through the air around me while, through my window, the moon, with human footprints on it, shines—well, I wonder what people will think of as "natural" about a century from now.

CHAPTER 7

Bugs on Reflection

I want to take a short break from microbes at this point, and shift the spotlight to the people who study them. Let us talk of what we think, what we know, and what we think we know. We'll be looking at the past twice, into the future once and, as a bonus, at a world that does not, strictly speaking, exist.

I'll begin, however, with a story that has all the necessary ingredients for a successful fairytale: a hero, a struggle, a flash of brilliance, and a frog.

Hole Punchers

Xenopus laevis (*X. laevis*) is a type of African frog that is widely used in developmental biology—a field of biology that tries to find out how a fertilized egg grows into an entire animal. Because its eggs are numerous, relatively huge, and grow very quickly, it's a highly valuable model organism for that field.

About twenty years ago, a researcher was studying the ovaries of these frogs and their influence on the embryos. The basic technique for obtaining the ovaries was simple: He would take a frog from its water tank, anesthetize it, cut open its abdomen, remove the ovaries, sew the frog back up, return it to its tank, and proceed to examine the ovaries. Then, one day, he asked himself a question that hadn't occurred to him (or to any of the other people who were doing similar experiments in labs all over the world): Why were the frogs still okay after such invasive surgery?

If you have to cut open a human's stomach, you take the patient into a sterile operating room; you work with sterile instruments—essentially, you take care not to allow any microbes to infect the open body, or they will cause inflammation and very serious problems, even death. You don't just slap them onto a table, cut them up, sew them back up, and chuck them into a dirty environment. How come these frogs were still happily hopping around in their mucky water tanks, then?

It seems obvious now, but it takes a special kind of genius to notice when things that are all right shouldn't be; and it takes a certain type of person to decide to switch over to a totally different field of research to try to find out why—and that's what this researcher did. The answer he found out is this: Frogs can secrete small molecules from their skin that act as antibiotics. These molecules have a positive electrical charge (like the positive side of a battery), so that when the molecule encounters a bacterium, it can enter the bacterium's cellular membrane, which is negatively charged; fold itself into a ring structure; and stay there. This basically creates a large hole in the bacterium's membrane, and really messes up its life (imagine puncturing a ball or balloon—it's a bit like that). The reason these molecules do not affect the frog's cells is because animals' cellular membranes are different from those of bacteria—they are not negatively charged. So *X. laevis* effectively sterilizes itself, and heals quickly, without further trouble.

This discovery opened up a whole new area of research: Up until then, antimicrobial molecules were something produced by fungi, molds, and plants—and by the bacteria themselves (for harming other species of bacteria); but no one had heard of animals producing them.[1] Sure enough, these small molecules were then found in many other animals, including humans—you have them protecting you in your saliva, your airways, your blood, and in many other places. When you were a baby, you got them from your mother's milk, which helped you fight bacterial infections from day one.

Several labs are now trying to come up with new varieties of this class of molecules to act as antibiotic medicine. With the antibiotics we use today becoming less and less useful, these small molecules may be humankind's saviors in the future.

It pays to pay attention to frogs.

Pasteurized History

In James Clavell's epic novel *Shogun,* which takes place in Japan four centuries ago, there is a memorable scene in which the recently arrived English navigator John Blackthorne is summoned before the Japanese lord Toranaga and questioned at length. At one point, Blackthorne tells Toranaga of contemporary European politics: The Netherlands, previously a vassal state of the Spanish king, has rebelled and is now at war with Spain. Toranaga, a ruler in a strictly hierarchical society, is furious—rebellion against a lawful king is

1. Not quite accurate, actually. Lysozyme, an antibacterial molecule found in human tears, was discovered by Alexander Fleming years before he had isolated penicillin, but its therapeutic potential was found to be limited. Today, it is used mostly in laboratories for isolating DNA from bacteria.

inexcusable. Blackthorne, whose head is in very real danger of rolling at this point, says, "But there are mitigating circumstances." Still, Toranaga will have none of it: "There are no 'mitigating circumstances' when it comes to rebellion against a sovereign lord," he exclaims, and Blackthorne replies, "Unless you win."

A tense moment: Blackthorne will surely die for his insufferable insolence, and the following thousand-or-so pages will have to be left blank. But, no, Toranaga laughs and says, "You named the one mitigating factor."

From fictional Japan, we move to the real France of the 1860s, where a battle raged between scientists, followed closely by the general public (or at least the upper classes, which are public enough for our present needs). On one side were those who backed the theory of spontaneous generation, which suggested that microorganisms were regularly being created from decaying organic matter. On the other side were those who claimed that life only comes from life—chief among them was the great Louis Pasteur, the man whose achievements rightly earned him the title "father of microbiology," alongside Robert Koch.

Pasteur had prepared sterilized yeast solutions in glass flasks that had necks that were twisted to such a degree that microorganisms could not, he rightly asserted, enter. Any life found there would have to have been spontaneously generated from the dead organic matter, he argued. Spontaneous generation did not occur, of course: Pasteur's swan-necked bottles stayed free of moldy growths for extended periods of time. These demonstrations, characteristic of Pasteur's virtuoso technical abilities and powerful logic, nearly closed the book on the debate.

I say nearly, because along came Felix Pouchet, Pasteur's main rival, who prepared a similar flask, only with a heat-sterilized hay infusion rather than a yeast-water solution. Mold appeared immediately. Pouchet could now argue that Pasteur's experiment proved nothing more than that spontaneous generation could not happen in the very specific case of yeast water. Here was some hard evidence that was very troublesome for Pasteur. It may also be troubling to you: How did life suddenly spring forth from a sterile container? Doesn't "sterile" mean that there is nothing living in there? So where did the mold come from?

The answer, we know today, is that Pouchet's hay infusion was not sterile: The heat-sterilization technique everyone was using at the time, the one we now call pasteurization, was adequate for getting rid of most microbial species, but no one knew then that some species could develop spores—those tough external shells that protect the microbe from adverse conditions, and can withstand very high temperatures. Those microbes that survive the pasteurization process can, once it is over, shed their tough external protective layer, become active, and multiply. They feast on the lactose in the milk, and excrete sour lactic acid as a waste product; this is why pasteurized milk goes sour after a week or so, even in the fridge. For preparing completely spore-free, sterile milk (the type we call long-life), an even more rigorous treatment is necessary.

We return to Pasteur, who was unaware of the existence of spores. What was he to do in the face of this challenge to his position? The scientifically correct thing to do would have been to try to replicate the experiment, and to then either find some technical fault with it (as he had successfully done with previous attempts by Pouchet), or come up with a theoretical explanation for Pouchet's results, which he could then test (perhaps suggesting the existence of the heat-resistant microbes that were indeed responsible for the result, or some other possible reason). Failing that, he could give up and concede defeat.

Pasteur did none of the three; instead, he all but ignored Pouchet's results. In this he was supported publicly, by favorable opinion, and privately, by his own convictions. Pasteur, a Catholic, was very much opposed to any type of thinking that had even a whiff of atheism, including the new wind of Darwin's theories that was blowing in from across the English Channel.[2] As a religious conservative, he maintained that life was created once only—by the Creator, at the beginning of time, as described in the scriptures—and would not, therefore, pop up on a daily basis. Pasteur's attitude fit in very well with the French political climate at that particular time, so he was not unduly pressed, and was able to shrug off his opponents.

Pouchet, who was a religious man himself and based his theories on (admittedly rather elaborate) religious grounds, was off the map. Pasteur won the debate and went on to further scientific triumphs. His victory, in this case, was not due to him being an exemplary scientist; it was heavily supported by fortunate public circumstance and by his preconceived notions.

There are two points from this episode that I found disconcerting, in an illuminating way: one, scientific; the other, historical.

The scientific point is that Pasteur conducted himself in a less-than-professional manner. We tend to view great men of science as embodying the spirit of science; as being dispassionate observers of nature who follow their observations wherever they may lead. This is an ideal and, as such, it's difficult for men of flesh and blood to embody it. A scientist, even a great one, is a human being, with all the human faults and fallacies this implies. What's more, men of great vision often tend to be hardheaded to the point of stubbornness. They

2. There is some debate over Pasteur's real piety: His church attendance was apparently somewhat lax; however, publicly, he always remained a true believer.

have hunches and convictions, and they also have their pride. For Pasteur to admit, after all his time and effort, that his rivals might have been right would not only have gone against his conception of nature, but would also have been quite a blow to his not-insignificant ego.

Pasteur was ultimately proven right, not by his own experiments, but by future researchers who followed his lead. This is a strange way of going about doing things, when you think about it, but that's how science works. It works because those who follow do not do so blindly, but check and recheck the ideas they receive from their predecessors, and then dispose of those ideas that are found to be inaccurate. Pasteur maintained that life comes from life. He was right. This is the verdict of science. Like the rebellious vassal, this is the one mitigating circumstance.

The historical point is that Pasteur, a great man of science, was opposed to Darwin's evolutionary ideas, and was fighting on the side of religious conservatism. It's a good lesson to absorb—that scientists are men and women of their times, and do not conform to future generations' notions and alignments. They work with the knowledge, understanding, and technical means that they have at their disposal, and they are also, in no small measure, influenced by their society. We should not mistakenly ascribe our own views to them, however large the role they performed in cementing those very views.

The verdict of history is complicated. How should we judge a man who was unprofessional, but accidentally right? Should we laud him for trusting his instincts, be disappointed at the way he gained the upper hand, or just be thankful that, for once, the truth won out (if by strange methods)?

It's your call. For myself, I wonder what Pasteur would have thought of the things we know today about life and microorganisms. Would he have been pleased and fascinated, or hurt by the modern, overtly secular conclusions that we've reached, based on his findings? *Shogun*, if I can return to it, is a book filled with strong men. Indeed, James Clavell seems to be of the school of thought which says that strong men make history.[3] Looking at Pasteur, I would perhaps add that if this is what strong men do, the history that they make is rarely the history they thought they were making.

The Box with the Pox

PANDAS FINALLY WIPED OUT

The ongoing effort to eradicate pandas has reached its conclusion.

"We've finally got them all," reported Thomas B. Splugg, leading panda research scientist and head of the international Panda Control Program. "These creatures shall no longer trouble humanity."

A globally coordinated endeavor involving extensive international measures and multimillion-dollar budgets has succeeded in reducing the worldwide panda population to only two living specimens, kept in isolation at ultra-secure facilities. While most authorities are content with this situation, there are safety-concerned activists who demand that the pandas be eradicated immediately, in order to prevent the possibility of one escaping into the wild.

"They're crafty little things," said R. L. Spangler, spokesman for PandaMonium, the Concerned Citizens Initiative. "We cannot risk even the slightest possibility of a future pandemic."

3. The opposing school of thought is that all men are but unwitting cogs in the inevitable march of history.

You'll probably never read anything like this in the papers. Even people who do not particularly like pandas would not rejoice in their extinction. It's true that environmental scientists and agencies sometimes hunt down certain species, but that's usually as part of an effort to preserve a specific local environment against foreign, introduced species (like rabbits and cane toads in Australia, the kudzu vines in the United States, and virtually anything that arrived in New Zealand after Captain Cook). Nobody wants to drive an entire species to extinction, though . . . unless, of course, it is a microbe.

In contrast to most biologists, microbiologists who study microbes that are harmful to humans see it as their goal to try to wipe out their research subject. Still, while medical science has been fighting infectious disease-causing microbes for centuries now, with all our efforts and all our discoveries and technology, how many diseases have been eradicated, never to return? How many species of bacteria or virus has humanity managed to bring to extinction? The answer is one.

Trying to eradicate a disease—an entire species of microbe—so that it never comes back is pretty hard. There are several obstacles that can prevent success. First of all, you need a reliable, effective vaccine for that microbe. And the disease also has to be limited to human carriers, otherwise you'll need to chase any other animals it can hide in, and either kill or vaccinate them all. Of all the infectious diseases, from the sniffles to AIDS, only one is gone—smallpox. Many other diseases, such as polio, no longer pose the threat they used to, but they're still out there.

A truly heroic international vaccination effort saw smallpox—a disease that had killed untold millions of people during some of the most horrific plagues in history—officially declared eradicated in 1979. Since then, no further cases of the disease have been recorded.

The last remaining samples of smallpox viruses, *Variola major* and *Variola minor*, are kept frozen in two secure labs, one in the United States and one in Siberia.

Why, you may ask, should we keep them at all? Why not destroy these last samples, and be done with it? We have the virus's genomic sequence, and we have the vaccines, so why should humanity keep one of its worst enemies frozen away in a box, like some bad horror-movie villain, where it could be stolen, or set loose by accident?

It's a good question with no clear answer. Some of the reluctance to dispose of it can be traced to simple procrastination (which has been very effective in staving off other horrors, such as doing the dishes), but more serious thinkers argue that we might not know what we are throwing away, and that future scientists may find out new things that we can't foresee today. What if the *Variola* virus still lurks, unsuspected, somewhere in deepest Africa or Greenland, waiting to emerge again in years to come? Even worse, what if, at some point in the past, a virus sample was taken from its lab, and is now in a terrorist's freezer, to be unleashed upon humanity when Armageddon is at hand? Granted, we still have the vaccine; but what if, for some reason, it's not enough?

It's difficult to measure all these what-ifs against one another. In the meantime, the pox boxes remain.

Come to think of it, the *Variola* virus is not really the only one that humanity has managed to exterminate. It's just the only one we have managed to exterminate *on purpose*. For every species of frog, mammal, insect, or plant that we inadvertently drive to extinction, there are myriad microbial species that have been living in and on it. What were they, and what did they do? What could we have discovered about them? We'll never know. They're gone.

Three Drinks and a Pump Handle (aka the Pettenkofer File)

Professor Barry Marshall of the University of Western Australia (UWA) received the Nobel Prize in Medicine in 2005. Professor Max von Pettenkofer committed suicide in 1901, aged eighty-seven, and today he is all but forgotten. What links these two very different people is their shared profession, personal courage and resolve, and at the heart of it all, a drink.

I'll tell you about Marshall first. I grew up during the 1980s, and I remember people saying things like, "I'm getting an ulcer from this job," to indicate how worried or overworked they were. The peptic ulcer was considered to be mainly a result of stress. People living stressful lives were producing too much stomach acid as a result of all that psychological pressure. These acids would eat away at the inner lining of the stomach, eventually creating a wound in it that was very difficult to heal. People were suffering, and although an antacid-based treatment was effective, it was usually temporary, and the ulcers would return within a year or two. Incidentally, this also meant that return business for doctors and the companies that made those treatments was quite lucrative.

Meanwhile, at UWA, doctors Robin Warren and Barry Marshall were finding something where nothing should have been: Our stomachs are so acidic that nothing was supposed to be able to survive in there, yet samples taken from ulcer sufferers' stomachs were found to contain a new microbe, which Marshall and Warren named *Helicobacter pylori* (*H. pylori*). They put forth the theory that *H. pylori* was, in fact, causing the ulcers.

You and I, who have been to the furthest reaches of the Earth together and have seen the inhospitable conditions in which microbes turn up, may not, by now, be overly surprised; but suggesting that ulcers are an infectious disease, rather than a stress-induced condition, constituted a radical shift in focus. The medical community was unconvinced at first. The evidence was inconclusive: Marshall and Warren could not show a causal link, the microbiologists were trying to butt into internal-medicine affairs. To top it all, they were from the back of beyond. I mean, *Perth,* for heaven's sake.

To gain respectability, a theory suggesting a causal relationship between a disease and a microbe should satisfy four criteria that are known as Koch's postulates, after the German scientist who devised them. Essentially, the criteria require you to show that the disease occurs when the microbe is present, and that it does not occur when it is not present.[4] Makes sense, doesn't it? The only problem Marshall had was that his tests (which were on pigs) were not working. At this point, Marshall decided to do something rather radical. Because he had not succeeded in testing his theory on animals, he turned to the only human body he could test it on: his own. He drank a culture of *H. pylori* microbes, and promptly developed an inflammation in his stomach lining (and, had his wife not insisted that he take antibiotics after two weeks, he might have developed a full-blown ulcer—probably the only case of a joyous ulcer in history). Marshall then had his colleague remove and analyze a sample of the lining, in which they found, as he had predicted, live *H. pylori.*

Marshall announced the results to the scientific community, and it caused quite a stir; indeed, the world finally stood up and took notice. Within a few years, *H. pylori* was declared the primary causative agent of peptic ulcers.[5] Two decades later, Marshall and Warren re-

4. There are some qualifications to this, such as cases of asymptomatic carriers. Thought you'd like to know.

5. To be fair, *H. pylori* is by no means the only cause of ulcers: Stress and other factors have their roles, too, though much smaller ones than previously maintained.

ceived the Nobel Prize for their discovery. Today, ulcers are treated with antibiotics (unheard of twenty years ago), and they usually pose no continuing problems.

Marshall's self-experimentation was courageous, outrageous, unethical, heroic, and several other things, too. It was not, however, unprecedented.

In the nineteenth century, when cholera was exacting a terrible toll on Europe, something similar was going on in Munich. Cholera, which had arrived from India in the early years of the century, was killing large numbers of people in unforeseeable outbreaks. Its causes were unknown and were, naturally, the focus of much speculation and heated debate. The European medical and scientific communities, which had hoped that the plagues which had ravaged the continent up until the seventeenth century were now a thing of the distant past, were suddenly faced with this new threat.

Many researchers and physicians believed that the disease was caused by miasmas: foul, disease-causing air. They were blamed for a large variety of illnesses—some public figures went so far as to suggest that even moral contamination could be spread through a miasma, and they proposed designing prisons so that this could be avoided. Supporters of the rival theory (that the disease was caused by a tiny waterborne life form) were not, initially, as influential.[6] Today, we know who was right and who was wrong; but, at the time, both competing theories were highly plausible.

In 1854, the British physician John Snow methodically analyzed the appearance and spread of cholera cases from an epidemic that hit central London, and tracked down its origin to a well pump on Broad

6. In fact, the common people were usually more wary of their drinking water than experts at the time thought they should be. Common sense is a fickle thing, but on this occasion it was eventually proven right.

Street. He persuaded the local authorities to remove the pump handle to prevent it from being used. Although the Broad Street epidemic was apparently past its climax when this happened, Snow's actions had a number of consequences: They helped protect the pump users from exposure to the contamination; they established the fundamentals of epidemiology, as it is, in essence, still practiced; and they provided some solid evidence for the germ theory of disease.

Indeed, Snow's controversial theory was vindicated: Within thirty years, Robert Koch had isolated and described *Vibrio cholerae* (*V. cholerae*), along with the causative agent of the disease, and had given the science of medical microbiology its foundations—his aforementioned postulates.[7]

Some people, however, were still not convinced. Enter Max von Pettenkofer, a respected German chemist and physician. Snow's results led Pettenkofer to investigate the matter of cholera, and he concluded that its cause lay in the moisture content of the soil. So convinced was he that a microbe cannot cause cholera by itself that, in 1892, at the age of seventy-four, he drank a sample of *V. cholerae* that Koch had sent him, just to prove that it would do no harm.

Are you shaking your head in sympathy over the poor man's fatally misdirected conviction? You've got another shake coming. Pettenkofer had the runs for a few days, but otherwise suffered no ill effects and was back to normal within a few days, having proved his point.

"What," you may ask, "is going on here? Didn't you just say cholera is caused by a germ? How did he not get sick?"

---

7. Interestingly, the microbe was, in fact, first isolated in 1854, the year of the pump; but the discoverer was Italian, and Italy was, at that time, leaning strongly to miasmatic views. So the discovery sank into oblivion until the time, the place, and the people's minds were ready for it.

No one knows. Sometimes people just stay healthy, regardless. Theories include suggestions that Pettenkofer was immune, due to a light case of cholera he had unwittingly contracted earlier in his life; that the sample Koch had sent was mostly dead bacteria by the time Pettenkofer had quaffed it; or that Pettenkofer's stomach acids (acids that normally inflict heavy losses on germs that enter the digestive system) were unusually strong and managed to kill off all the germs. Whatever the cause, the fact remains that Pettenkofer and miasmatic theories were dramatically vindicated in the eyes of his supporters. We now know that it was a lucky fluke; but, at the time, it must have seemed like quite a proof.

Although this was something of a setback for germ theory and medical progress, it was not that big an obstacle. Germ theory was on its way to becoming firmly established, and little harm was done. In fact, Pettenkofer's efforts, like those of many supporters of miasmatic theories, turned out to aid public health in many respects, as they led to improved hygiene and sanitation measures in European cities. They had the wrong idea, but they led to a good outcome. Progress sometimes works that way.

The bigger issue here lies in the concept of proof. If we contrast Pettenkofer's dramatic act with John Snow's careful analyses, we may glimpse a hint of what science is all about. Science is not usually impressed by one-offs. A spectacular action may direct initial attention, or provoke sympathy, toward a certain spot or stance. But, ideally, it is the considered (and, above all, the quantifiable) that ultimately prevails—all the more so when dealing with complex systems such as human bodies and populations, which is why epidemiology is such a necessary and useful area of study.

We shouldn't be too hasty in lauding or scorning, though; both Snow and Pettenkofer had good reasons for thinking they were right—and their theories were sound, as far as the then-current evidence was concerned. The fact that we know in hindsight who was right and who wasn't does not necessarily make one man heroic and the other ridiculous.

This seems like such a morally confusing tale, doesn't it? People doing the right things nevertheless have the wrong theories, while people doing the wrong things receive recognition and honor. It is therefore comforting, in a somewhat macabre way, to tell a third story that has a different outcome.

Cholera epidemics had their political side, too: Outbreaks were overwhelmingly prevalent in the poorer quarters of the cities. Wealthy landlords and water-company owners naturally had a tendency to prefer to believe the miasmatic theory, because water systems were costly to maintain and improve, whereas air quality was, at that time, not their financial concern. One landlord who had received incessant complaints from his tenants about their water supply being contaminated by cesspool water went down himself, sniffed his tenants' drinking-water supply, pronounced it to be fine, and demonstrated his conviction by drinking a glass of it. He died of cholera a few days later.

A hundred years later, disease, money, and politics are still intertwined. AIDS is a current leading example of this relationship, though there are many others that are just as fraught. Cholera and other waterborne illnesses, though relatively easy to treat (with antibiotics), and even easier to prevent (with clean water), still cause much suffering. Much like that ill-fated landlord, we of the developed world (myself not excluded) still tend to turn a blind eye to what goes on in the poorer quarters of our world.

As for the peptic ulcer, I find the story uplifting for a different reason than you may suspect. Some people see it as a shining example of one man's selfless struggle and ultimate triumph against a conservative, unheeding medical community controlled by ruthless pharmaceutical companies interested only in making a profit by selling expensive drugs to patients. I think that the truth is (as usual) more subtle.

Marshall's drinking session received much interest in the years immediately following. He was not shrugged off as a quack: A number of researchers set out to disprove his findings, found them accurate, and truthfully reported so. Within little more than a decade, the entire mode of medical thought and practice on this subject shifted totally, from the traditional view to the new revolutionary one. New treatments replaced traditional ones, even though they had been a highly profitable source of income for specialists and the drug industry.

Big ships change direction slowly. I think that the medical establishment moved quite quickly in the case of the peptic ulcer, especially if you consider the sizable shift in perception and application that was called for, and the money involved. Without reducing Professor Marshall's achievement by even the least tiny bit, when one considers the trials and tribulations that past revolutionaries have gone through, one decade and some stomach trouble is not the highest price some people have paid for their opinions.

Two well-meaning germ drinkers, a century apart, believed so firmly in their findings that they were prepared to undergo the most personal and hazardous of tests. One drank his microbe broth in order to prove it could cause illness; the other drank to prove it couldn't. Both efforts were considered successful; one was later proven to be utterly wrong; and the other is, as far as we can tell at this point, right. Is there any lesson here? Perhaps this: In science, dramatics and heroics may help you get your theory's foot in the door but, in the long run, they will do no more than that. At some point, hopefully early on, a theory will be considered on the merits of its accuracy, not on its presentation. This is the core of the scientific method.

99.9 Percent

I recently went on a picnic with some friends by the Yarra river in my hometown of Melbourne, Australia. After the munching had subsided, I took one of them for a stroll and showed him the hundreds of large bats hanging upside down like weird black fruit from the gum trees along the riverbank. My friend was suitably impressed: "Dwagons," he observed.

After I had safely deposited the budding zoologist in his daddy's loving arms, I got to thinking: Why couldn't they be dragons, as far as he's concerned? Here is a creature he has never before encountered in the flesh that's clearly not a bird, but that's just as clearly flying around on strange-looking leathery wings. In the three years of my young friend's life, up until now, he had been told stories about dragons, shown pictures of dragons, and watched films starring dragons. So why would he doubt the existence of a real-life dragon, if a suitable suspect presented itself?

How strange, I mused—while, behind me, naptime negotiations were escalating into an all-out bawlfest—are the boundaries of our knowledge. They encompass hordes of mythical beasts that inhabit only our imaginations, from the troll and the werewolf to the honest politician, and yet we remain totally unaware of countless living things that do exist, for the trivial reason that they are too small for us to see.

When microbiologists wish to study a microbe found in a sample taken from nature, traditionally they will first try to isolate it by diluting their sample until only one microbe per culture is left. It's important that they do this, because it's very easy to accidentally get two different species of microbes on the same plate—when that happens, the entire analysis goes awry.

After making sure there is only one species in the culture, they allow it to multiply until they have a large enough quantity, at which stage they try to characterize the organism. What shape and color are the colonies it forms? What nutrients does it need? Under what conditions does it prosper? Then they analyze the microbe's DNA, its proteins, and its metabolism. It's a lengthy, painstaking process and, what's more, it shows no sign of ending. As we've seen again and again in previous chapters, even *E. coli*, the most comprehensively studied organism in the world, is still surprising researchers, more than a century after its discovery and a decade since its entire genome was sequenced. There is no organism in the world about which we can say that we know all there is to know.

But the problem of knowing things about microbes only starts there, because this method of studying microbes depends on us being able to grow them in a culture, and it now appears that the microbes that deign to grow in our laboratories are the exception, rather than the rule. For every species we have been able to isolate, there are hundreds, perhaps thousands, which we cannot isolate. Fittingly, we call these uncultivatable microbes. We can't even know how many we're missing—how can we count something that we can't identify?

Cleverly, that's how. Here's a method for finding out things about creatures that we can't pin down: You take a sample out of whatever environment you wish to study—soil, seawater, a human throat, anything.[8] In each sample, there are countless different species of microbes. You don't isolate these microbes; rather, you lyse them all (burst them like bubbles), extract the DNA from the entire sample,

8. Even a teaspoon-size sample can contain millions of bacteria of heaven-knows-how-many kinds—and that's not counting the viruses.

put it together, and sequence it. What you get is a very large number of fragmented sequences.

There is, apparently, a lot to be learned from this chaotic jumble of genetic information but, in order to make some sense of it, we need to pass it through some very powerful computers. These computers use programs that take a fragment and analyze it by comparing it to known sequences that we have found in other organisms in the past, or by analyzing the sequence and giving educated guesses about the genes' functions. Another method compares two fragments, looks for any overlaps and, if some are found, draws the tentative conclusion that they are part of the same sequence. Multiply this process by a lot and there arise some probable whole genomes (it can be likened to assembling several puzzles whose pieces are all jumbled together at the start). Once we've constructed several separate genomes, we can try to analyze them and understand a little of what these microbes may be like, what they do, and how they live.

All of this takes serious computing muscle.[9] Still, the results are worth it: New, never-before-seen genes are being found by the bucket load (literally so—the researchers go out to sea in boats, using buckets to obtain these water samples), and the stream of new, and very possibly useful, information is constantly increasing. It turns out, for instance, that relatively close regions of a certain sea, which look just the same to us, can have very different microbial inhabitants. Why is that? Why are these guys all bunched up over *here*, while those guys are huddled way over *there*? What do they know that we don't?

If that doesn't particularly rock your boat, consider this: A multitude of new genes is being discovered—genes which code for proteins

9. Which is one reason why these techniques are only now being put into action: Ten years ago, this was simply too much data for a computer to handle.

that may have all sorts of novel practical uses, in anything from industrial processes to new medicines or better recycling processes. These scientists are, quite literally, fishing for information, and they're making some promising catches. Craig Venter's yacht, the *Sorcerer II*, returned from its round-the-world expedition in 2007 bearing news of dozens of previously unknown microbe species and a jungle of DNA sequences—a tentative start to a Genome Project spanning entire oceans. The data are currently being analyzed, and within a few more years, stories of these newfound microbes will most likely emerge. With a bit of luck, you might hear about them on the news.

What scientists have now is still just raw information; but a great deal of information, when correctly handled, can sometimes turn into a measure of knowledge and, on occasion, even a bit of wisdom.

Besides, you never know what will be pulled from the ocean depths. Personally, I'm hoping for very, very small dragons.

BONUS TRACK #7

Are You Game?

In September 2005, the massive multiplayer online game World of Warcraft experienced an unforeseen calamity: For a few days, players' characters were dying in Black Death–like proportions.[10] It wasn't the first time this kind of thing had happened in the gaming world, but this particular incident proved to be quite instructive.

The game's administrators had introduced a new dungeon in which an illness called corrupted blood could be caught by the virtual character, either if the character was hit by a particular spell from its source—one Hakkar the Soulflayer, a Blood God by vocation—or if they caught it from other characters within the game. The new dungeon was supposed to be restricted only to high-level players who could withstand the damage done by the disease, but the virtual virus escaped (some say that malicious players smuggled it out, in order to purposefully harm others) and started infecting like mad. Within hours, entire virtual worlds of some game servers were infected, thousands of bodies were strewn in the streets of the virtual cities, and chaos reigned. After some initial bumbling by the administrators, the infected game servers were rebooted, and order was restored. Would that it were only so simple in the real world.

Although it was a pretty messy error, it did have some beneficial effects, especially for epidemiologists, who saw this as a useful way to gather important data.

When epidemiologists prepare for future epidemics of infectious diseases, they devise mathematical models that try to simulate how people will react in the case of a disease outbreak. They take into account historical cases, the population's psychological factors, and its relevant physical attributes (population size, density, transport, and infection routes, for example). The problem is that, in historical cases, large chunks of information are always missing, and psychological factors are notoriously hard to replicate. A disease outbreak is a chaotic event that evokes highly unpredictable reactions from individual people. How does one mathematically simulate mortal fear or confusion? It's not easy.

Studies that place volunteers in simulated epidemic situations and test their reactions can help, but asking someone how he would react if he were in a state of utter panic does not necessarily elicit a very

---

10. Crash course for the unenlightened: World of Warcraft is a game involving a virtual fantasy world, complete with geography, rules, and strange beings. A player controls a character, or avatar, and carries out quests and missions, or does whatever he or she wants. Players interact with each other, gain possessions, and advance in levels. Millions play it daily.

reliable response. A volunteer in a simulation always knows (somewhere at the back of her mind) that she can walk away if she wants to. The models are just not real enough.

In a virtual gaming world, researchers have, to some extent, a unique opportunity to overcome those difficulties: They combine a simulated environment (in which the data is all there and very reliable) with thousands of actual human participants who are reacting in real time to the situation and to each other; and, crucially, in a game such as World of Warcraft, the players develop an emotional attachment to the characters they play. They invest long hours, much attention, and often real money on their avatars. They grow to care about their character (and about fellow characters), sometimes quite deeply; as a result, their reaction to a threat is more natural, reflecting true crisis behavior.

The players in the World of Warcraft epidemic behaved in many different ways, all of them very human, but some never before encountered in a simulated model. Some player characters were slow to catch on and died, many ran away, and some stayed behind selflessly to help others.[11] A small minority also behaved maliciously by knowingly infecting others, while some were opportunistic, seeking to profit from the situation. A number were also just plain stupid, coming into the quarantined zone just to see what was going on, and dying. In short, people were being people.

The characteristics of this event were illuminating: High-level players could teleport to distant corners of the game world within a very short time—something like a virtual version of modern air transport—and the disease was apparently smuggled out of its restricted zone, which gave the whole thing a distinct flavor of bioterrorism. The authorities were slow to react, and nonplayer characters could be asymptomatic carriers—they could catch the disease and pass it on, but not be affected.

These accidental properties were reminiscent of past epidemics (especially the SARS epidemic of a few years ago) and a cautionary glimpse of possible future ones.

Of course, a virtual epidemic is not a perfect replication of a real one: Reactions are sure to be quite different, because characters can be resurrected—and in any case, a virtual life is hardly as important as a real life.[12] Still, there may be some benefit for science in running other better-planned outbreaks. Gaming companies have refused, so far, to

continues

---

11. These noble actions actually contributed to the spread of the disease. Good intentions don't always achieve the desired effect.

12. Except for those who spend virtually their entire lives on these games. Such people are sometimes, paradoxically, encouraged to "get a life," as if they don't have enough of them to manage already.

unleash large-scale infections on their customers, pointing out that it's not fair to them. I would suggest that life isn't supposed to be totally fair, and that a dose of randomness and unpredictability is precisely the ingredient needed to achieve the state of realism that they're working so hard to create—especially if you consider that real-life war casualties throughout history were mostly due to illness, rather than to direct combat. Players in World of Warcraft didn't know that the outbreak was unplanned, and many thought it was cool to have such a realistic feature introduced into the game. The game's development team is apparently onto this, but I doubt whether they'll be handing the reins to the epidemiologists anytime soon.

In the end, scientists may have to build their own games, in order to attract volunteers and carry out scenarios. The problem is how to build a game so it will attract volunteers who won't know what they're volunteering for, in order to maintain the crucial element of surprise and the required candid response. A tricky task, no doubt, but it can be done. Besides, teenagers glued to their consoles all over the world will now have the perfect response to their mother's complaints: "But, Mom, it's for *science*!" Try arguing with that.

On the bright side, there was one group that benefited from this inadvertent plague. Because it affected the human-player characters, the gamers who played *orc* characters (who were in no danger of catching the disease) were having the time of their lives. "We just laughed and laughed," reported one.

Game on.

CHAPTER 8

Bugs Encore

Popular-science writing often makes complex ideas seem a lot clearer and more organized than they are. That's the ultimate objective of writers—to explain things. But we need to beware of making it all seem too tidy. Take genes, for example: This gene does this job, an article will explain, and that gene affects that trait. How nice. But when we look at an actual genome, all notions of tidiness are quickly dispelled. At first sight, a genome looks seriously messed up—genes are jumbled all over the place, overlapping each other, with small parts stuck in unexpected places, while completely unrelated genes are bunched up together, and haphazard DNA sequences are repeated for no apparent reason. Nature's notions of organization and order seem to be rather different to how we humans understand these terms: If people (my mother, for example) had been in charge of this kind of project from the start, a genome would be properly sorted out, with genes arranged in sectors, labeled for future reference, and maybe even color coded.

The real state of nature always reminds me of a story about an engineer that a friend of mine once told me: He and the engineer were working for a major electronics company, and the engineer was undoubtedly a genius in the field. His designs were amazing and hugely complex, but they worked, always either achieving or exceeding specifications. No one knew how. I was impressed to hear this. "But what happens," my friend admonished, "when he leaves, or even gets sick, and something goes wrong with the product? Or if we want to develop the project further, make a newer model? Nobody can figure out what's going on in there! It'll all be useless once he's gone."

In a way, that's what nature's like, except for the designing bit: If you look closely enough, you see that natural systems aren't designed. They grow—usually inside, attached to, on top of, into, or out of each other, like tubing in plumber hell. Nature works, that's for sure, but it does so according to its own schemes, and we have the job of trying to understand how, if we want to fix or improve anything.

In this final chapter, I'm going to talk about some cases of untidiness in nature: about things inside other things, and things affecting—and even becoming—parts of other things. In other words, I want to examine boundaries. The first of these is one we usually take for granted, until we inspect it closely: The boundary between life and not-life. It can be a bit of a trial at times.

Life Sentences

Hear ye, hear ye—all rise! The court is now in session, the honorable Judge You presiding. In this sitting, the court will review four separate cases involving entities that claim to be alive. Formal definitions differ, so you may decide for yourself whether each entity should be considered a living organism or not.

Case One: Silent Majority

Viruses of all kinds are the most common biological forms on Earth, easily outnumbering all the rest of us combined. Anywhere that life is found, viruses are sure to be there, too.

They have genes, they have proteins, they have life cycles, and they evolve. The problem is that they have no metabolism and no independent existence: They do not breathe or eat and, outside the host cells that they infect, they simply float around and wait for the next host cell to turn up. When they do infect a host, they hijack its machinery for their reproductive processes.

Some say that, if they have genes, evolution, and reproduction, they have every right to be treated as a living thing, even if they are parasitic. Aren't there many other sorts of parasites, some much bigger, that are definitely considered to be alive? Why discriminate against viruses, just because they're small and lack a few qualities?

Others argue that metabolism and some sort of basic independence are crucial to the definition of life, and that viruses are no more than a few genes surrounded by a protein shell. It would be absurd to award this simple piece of matter the honorary title of "living"—it would cause the definition to lose any descriptive value it may have had.

Case Two: Jumpin' Jeepers

The problem is intensified when we look at transposons. These wee bits of DNA are found inside most genomes, including our own. At its most basic, a transposon is a single gene flanked by two small sequences of DNA, each a mirror image of the other. The gene codes for one protein, whose function is to attach to the flanking sequences, cut out the entire transposon, and move it to another spot in the genome.

There are many variations on this basic theme: Some transposons move around a lot, some hardly at all. Sometimes they duplicate themselves, and multiple copies of a single transposon appear in one cell. Some can insert themselves anywhere; others are more finicky about the DNA sequence they inhabit. There's also a variation in transposon size: Very often we find other genes joining the first gene, in between the flanking sequences, to go along for the ride.

When these jumping genes were first discovered, back in the 1950s, they caused much confusion in biological circles. Genomes were considered to be fixed, stable things and were not supposed to change every once in a while. Nevertheless, to the dismay of many biologists who thought genomes were nice, sensible, easily studied things, this disorderly hopping around proved to be exactly what happens, and quite frequently, too.

There's a debate about the role transposons play in the organism. What's the use of them? What do they contribute to the survival of the cell they inhabit? A crucial point here is that sometimes transposons are seen to do harm. They hop in right into the middle of an important gene and disrupt it completely, which doesn't help anything. What possible benefit do they confer that is so great that evolution opts to keep them in the cell?

There are several possible answers to this question, and they may all be correct at once. One explanation lists various advantages that cells and organisms gain by transposon activity; for example, bacterial transposons sometimes carry genes for antibiotic resistance inside themselves and, when they move from bacterial cell to bacterial cell, they make the accepting cells resistant, which can be very advantageous to the accepting cell. Thus cells gain more flexibility and can adapt more quickly to change, but at the cost of an occasional danger or disruption to their genome. A very different kind of explanation suggests that the question "What good is it to the cell?" is the wrong sort of question entirely. The right way of looking at it would be to view the transposon itself as a separate entity—a genomic parasite, selfishly living and multiplying inside its host genome. All the variations between them, described above, are simply survival strategies. Sometimes, this parasite finds it in its interests to help its host along and cooperate with the genes around it, and sometimes it doesn't.

One gene with self-interest? Not unheard of and, indeed, a good example of the "selfish gene" evolutionary theory; but, if it is capable of self-interest, individual multiplication, movement between hosts, and evolution, should it be considered at least as alive as a virus?

Case Three: A Maddening Tale

If a transposon, which essentially consists of a single gene with an attitude, can be thought of as "alive" in a certain (albeit limited) respect, how about an entity that doesn't contain any genes at all?

One of the world's most intriguing mysteries first appeared in Papua New Guinea (P.N.G.). It manifested as a disease called kuru, which affected members of a certain P.N.G. tribe, causing them to gradually go insane and then die. Only certain tribes manifested the disease, while nearby tribes remained unaffected. A similar phenomenon called Creutzfeldt-Jakob disease (CJD), named after the researchers who first described it, had occasionally but very rarely popped up elsewhere in the world. When examined postmortem, the brain of the patient suffering from CJD would resemble a sponge.

What caused it, and why it was so prevalent in certain P.N.G. tribes and nowhere else, no one knew. The most reasonable idea was that it had a genetic basis, but the outbreaks in the tribes looked more like the result of an infectious disease. A similar disease known as scrapie had been seen in sheep, and another more famous disease of this sort, bovine spongiform encelopathy (the infamous mad cow disease), was responsible for more than 160 human deaths in Britain, generating much alarm in the Western world.

The cause of these diseases has now been traced to prions, normal proteins within our bodies that appear in an abnormal, dangerous form, with the protein kinked in a particular way. The weird, nearly unbelievable thing about it is that when a kinked protein meets an unkinked normal protein of the same sort, it kinks the normal protein into exactly the same form. You can see where this is going: A single kinked protein can "convert" another, and they both convert others, and then some more, until there are huge amounts of them. They tend to clump together in aggregates.

The prions can move between cells, and they can also move between hosts. This is exactly what happened with kuru—members of the tribe that suffered from the disease had been engaging in a ritual in which they would ceremonially eat the brains of a dead tribe member. If one of these brains contained prions (the first case may have been a pure genetic accident), they could be transferred to those who ate it, who would then, in time, also die. This did not happen every time—presumably, the prions wouldn't always make it into the new host's brain—but, when it did, death would follow, and the cycle would continue.

Cows, on the other hand, are not known cannibals. How did a disease that seems to be infectious only to brain eaters come to be an epidemic among cows? What horrible things are those innocent-looking bovines up to when we're not looking?

The answer is, it wasn't them—it was us. In the interests of saving money and increasing profit, modern farming practice was to take all the bits of the cow that were left unused by the butchers, grind them up together to make a cheap, protein-rich mixture, and add it to the cattle feed as a nutritional supplement. These supplements were, of course, subjected to sterilization procedures to prevent infection; but prions are proteins, not bacteria or viruses, and sterilization does not affect them—a fact no one suspected until it was too late. Cows were regularly made to eat other cows' remains, then prions found in one dead cow's nerve cells were transmitted to other cows, and humans then caught it from eating infected beef.

The ethics and epidemiology of all this are not our concern here, important as they undoubtedly are; what I want to focus on is the fact that here is an infectious-disease-causing agent that can spread between hosts, and even different species, but is neither a bacterium nor a virus. In fact, it contains no genetic material whatsoever. It's a protein, nothing more—and yet it functions very much like any other parasite.

Can it be called living? Can anything be called alive if it has no genes? Is it just a normal cellular function gone wrong? Or can the protein itself be called a gene, albeit a very irregular one?

Case Four: Poor Devils

Tasmanian conservation workers are very worried, and for good reason: There's an epidemic raging through the Tasmanian devil population. These strange, unique creatures, the largest surviving marsupial carnivores, which are found nowhere else on Earth, are rapidly succumbing to a disease called Tasmanian devil facial tumor.

Tumor. Cancer. But this is a book about microbes, and a tumor is not a microbe. Why am I talking about cancer here?

I'm including this disease here because it is a unique example of a transmissible cancer: It moves from one devil to another. The tumor cells spread via bites that the devils inflict on each other as a part of their mating behavior (this is one really intense little bugger). The cancer then spreads through the body, and the devil usually dies within a few months.

Cancer is not a contagious disease. We do not catch it from other people. There are some cancer-causing viruses known, but they operate in a very different way—they occasionally disrupt the host's genes in such a way that their anticancer defenses are compromised; but for the tumor cell itself to pass between hosts is unheard of.

There are factors that contribute to the uniqueness of this case: Most animals do not regularly bite each other in the face as a means of social interaction, and apparently, Tassie devils are so inbred (due to their population being small, even before the epidemic hit) that there's not enough variation in their immune systems to fight the tumor. They're all too alike—similar enough to all be susceptible to the same type of cancer. Be that as it may, it still seems a very improbable disease, and we needn't be concerned for our own safety: The chances of such a thing happening in humans are very small.

Saddened as I am, I still find the Tassie devil situation fascinating. Tumor cells are accidents of the body. They happen when a cell inside a body loses its brakes and starts multiplying without inhibition. In fact, it happens all the time—precancerous events occur inside a human body on a daily basis—but a healthy immune system takes care of nearly all of these cases. On the rare occasion, a tumor becomes uncontrollable, spreading from its original position into other parts of a person or animal, causing illness and (provided modern medical science does not prevail) death. Here the matter rests—the patient either dies or gets better but, in any case, the cancer is gone. This is not a viable, long-term survival option for the tumor-cell lineage. But in the case of the devil, it is: Tumor cells have become transmissible.

This disease, originating from one single cell, has now spread all over the Tassie devil population. All of these tumors in all of these devils are, in effect, a single, widely dispersed colony.

All life was once single-celled. During evolution, some cells banded together and formed "cooperatives." The Tassie devil tumor cell has, in a sense, come full circle, reverting back to unicellular form. How very antisocial of it. We can argue that all tumor cells do that, and that the only real difference between this tumor and any other is that this one has managed to infect a (very homogenous) population, rather than one individual. Once it goes through all the devils, though, it too will reach its dead end.

That much is true; nevertheless, this is the first time I've heard of such a thing happening.[1] Unlike all the other cases I have brought before you here, this is a case that definitely involves a living cell—but can it be said to be independently alive? Should we call it an infectious agent? Can we think of it in the same terms that we think of a pathogenic bacterium, even though it's a mammalian cell? Is it an organism now?

You be the judge.

Junkyard Genome

The Human Genome Project (HGP) was a global research project that set out to identify all the genes of the human species. It was estimated that, when it was underway, the HGP would map out about 100,000 genes. As it turned out, there were a lot fewer genes discovered—less than 30,000. This discrepancy in our expectations brings to light all sorts of questions. What do we mean when we say

1. While researching this, I discovered that it's not a unique, bizarre episode, after all: A nonfatal tumor called Sticker's sarcoma, found in dogs, appears to behave along the same lines. Who knows how many of these things there are out there? As a precaution, until medical science comes up with some answers, I've decided not to allow anyone to bite me in the face.

"gene" (which I discussed briefly, way back in the introduction)? And what caused this overestimation in the first place? Partly, the answers relate to microbial genes in the human body, which are found in two places: inside the human genome, or inside microbes living on, in, off, and with humans.

The human genome contains many bits and pieces. We know, or have a fair idea, what around 2 percent of these do—the rest are still under investigation. Because they don't appear to serve any useful function, they were initially known as junk DNA. That term is not very popular now, because the more scientists look, the more functions they find, or guess at, for these apparently useless batches of DNA that we carry around in every cell of our bodies.

Some of the bits and pieces are probably doing something useful, only we can't tell what that is yet. There are most likely plenty of crucial regulatory sequences in there—researchers recently found a whole new genetic regulatory level, called RNA interference, which we hadn't known about before, and it's fair to assume this is not the end of the story.

Another fraction of the genome consists of all sorts of short sequences and defunct genes: bits and pieces that we are not usually bothered with, and that may be a bit of a nuisance at times, but that are occasionally a handy source for much-needed components.

Still, a large fraction of our DNA is, as far as we currently know, parasitic. Over half of our genome is composed of transposons of various kinds. In fact, a huge swathe of our genome—about 10 percent of it—is just hundreds of thousands of a repeated, single, short sequence known as *Alu*.

Many of the transposons do not care, molecularly speaking, about the human race's benefit, and seek only to reproduce themselves. If we add to that some nonfunctioning viral DNA that is left over from past infections, we can see that our genome—so often described as "a blueprint of a human"—is, to a large extent, a collection of wee selfish bastards who've come along for the ride.

Nevertheless, some of these repeating sequences do seem to have a job to do. Transposon-like sequences are heavily integrated into our genome's regulatory networks, and recent genomic studies show that there are hundreds of further transposon-like sequences that are nearly identical in many species. This hints strongly that their yet-unknown role may be an important one. They may have come along for the ride, but they've exploited the system long enough to have become an integral part of it.

All Over the Place

Less than 30,000 different genes in the human genome? Yes. Less than 30,000 different genes in a human body? Most certainly not. Outside the genome, outside our cells, there is a wealth of genetic variety that we've barely begun to explore. For every cell in our bodies, there are ten bacterial cells in and on it, most of them inhabiting the intestines. It is estimated that microbial genes outnumber our genes by about a hundred to one. Microbes are infesting every bit of us that's not sealed off to them.

If reading this makes you want to jump up and head for the shower or immerse yourself in disinfectant, calm down: Only a very small fraction of these microbes are harmful, and our immune system generally keeps them under control. Hygiene is very commendable, up to a point: All the ads on TV that say, "Use X! Get rid of household germs!" are largely panicmongering, unless you've got infants or severely ill people living in the house. More common sense and less spraying of nasty chemicals is the rule of thumb.

Most of the microbes that live with us are positioned somewhere along the spectrum ranging from "mildly annoying" to "positively essential." In Chapter 6, we saw how ruminants rely on their gut bacteria to digest their food for them; much the same thing goes on inside our own digestive system, with a diverse host of microbes (most of them bacteria) breaking down various foodstuffs for us, and engaging in an intricate dialogue with our immune system, helping it to develop properly and to prevent harmful pathogens from settling in. These actions are, first and foremost, in the microbe's own self-interest, of course—that's what symbiosis is: two sides that each profit from their relationship.

How do gut bacteria first enter the gut, though? A fetus inside the womb is sterile, and the initial transfer of microorganisms occurs during or shortly after birth.[2] Eventually, in the first year or two of a child's life, they gradually build up stable, thriving gut flora.

Interestingly, gut microbes have recently been linked to obesity. Obese people and lean people have been found to have a noticeably different composition of microbial populations in their intestines, and there is some initial evidence that this difference may be a cause of obesity, rather than an effect of it. Obesity is obviously a result of many different factors, and there's a long way to go with this sort of research, so I wouldn't suggest relying on microbiology to make you supermodel-size just yet—still, remembering Marshall and his ulcers from the last chapter, we can at least say that this type of revolution in medical thinking would not be unprecedented.

Gut microbes have been tentatively implicated in the contribution of other types of health problems, too, from diabetes to tumors. Our understanding of the role that microbes play in these illnesses is

---

2. Note that I have tastefully refrained from using the words *contamination, mother,* and *feces* in this description, so as not to cause too much unease.

still far from complete. We also cannot yet understand what, if anything, the microbes themselves stand to gain from the disease. Theories range from simple backstabbing opportunism on the microbes' part to an imbalance triggered by improper diets or antibiotic abuse—human practices that have been steadily on the rise in the past few decades.

I hope the point is clear: Microbes are an integral part of human health. A germ-free person would be a severely sick person. The interactions between our resident microbes and us are, in all probability, as complex as the interactions among our own cells, and the boundaries are becoming more diffuse the more we learn. Our microbes are as necessary a part of our metabolism and our development as we are of theirs. Every human body is really a whole mobile ecology, complete with niches, competition, population dynamics, and environmental influences.

In late 2007, the U.S. National Institutes of Health announced the beginning of the Human Microbiome Project, a large-scale venture to sequence all human-associated microbial DNA. The information gathered by this project should enable us to start mapping out the relationships between man and his resident microbes. I'm eager to hear what they find, but I'll have to be patient; as of April 2009, the project is mapping the sequences of over seven hundred microbial species—and none of these sequences is at present complete. So, we wait. I think it'll be worth it. Call it a gut feeling.

The Inner Them

There is one category of DNA found in our bodies that is neither strictly human nor strictly microbial. It has been used to chronicle our human ancestry, but it turns out that it has an even deeper tale to tell.

Mitochondria are small structures found in nearly every cell of our (and every other animal's) body. They function mainly as the "power plants" of the cell, generating from its food supply the usable energy that the cell requires.

Mitochondria contain DNA (which is initially surprising, as the rest of our DNA is concentrated inside the cell's nuclei, not scattered around any old where) and, unlike the rest of our genome, which is a mixture of both our parents' genomes, it is inherited exclusively from the mother. My version of mitochondrial DNA was passed on to me from my mother, who got it from her mother; but I, being male, will not be passing it on to my children. This unique pattern is very useful for constructing genetic lineages and for studying human genetic diversity, because it can be traced back reliably for many generations.[3]

The point at which microbes enter this story is a rather distant one. About 2 billion years ago, one ancient microbe was swallowed by another. This usually implies that either the swallower ate the swallowed whole, or else the swallowed parasitized the swallower (as we have seen with *Bdellovibrio,* whose ancient relative may well have been this swallowed microbe). In this case, however, a sort of truce ensued—probably a very uneasy one, at first, if not a positively hostile one. Over the generations, however, these two very different organisms began to adapt to each other, and even began to rely on each other's abilities.

This state of affairs, known as endosymbiosis, gave this newly formed partnership a clear advantage over regular organisms: They

3. One such study resulted in researchers naming the matrilineal most recent common ancestor for all currently living humans—they called her mitochondrial Eve. This tongue-in-cheek term subsequently gave rise to much creative misunderstanding in the popular media.

could exploit multiple resources and work more efficiently. As a result, their numbers grew. This double act was the precursor to every eukaryotic (nucleated) cell—the type of cell that makes up all multicellular organisms. The "swallowed" microbe has degenerated in the eons that have passed since that time, becoming the fully integrated battery that is the modern mitochondrion.

As far as the mitochondrion is concerned, it has certainly struck a good bargain. It has exchanged freedom and autonomy for safety, cooperation, and a better chance to proliferate. What more could a microbe wish for?

It might seem that this ancient union was a freak occurrence, and that we should thank our lucky stars that it came about, but evidence indicates otherwise: Other organelles have proven themselves, pretty conclusively, to have once been free-living microbes—chief among them, the ultra-important chloroplast, the organelle responsible for photosynthesis in all plant life. This process has also been seen in real time in the lab, when an amoeba was "persuaded" to retain an infecting bacterium and, later, could not survive without it. Recently, an algal species called *Hatena* (Japanese for "enigmatic") was found to be in the midst of an evolutionary endosymbiotic process: When a *Hatena* cell containing a certain photosynthetic bacterial cell replicates, one daughter cell gets the bacterium and carries on living and photosynthesizing, and the other, devoid of bacteria, has to seek and swallow a new bacterium from its environment. In time, this will probably stabilize into a more sensible arrangement, and *Hatena* will no longer have to undertake this ridiculous method of having to hunt for its own inner organs while they try to wriggle away.

Some other variations on this theme include cases of multiple endosymbiosis (a microbe within a microbe within an insect); and the case of one type of sea slug that robs algae of their chloroplasts and installs them on its skin, creating the world's only solar-powered slug. What will they think of next?

All these goings on are highly confusing—at every level, everyone seems to be composed of bits of everyone else. Thankfully, some things remain outside the jurisdiction of microbes. For instance, microbes can't determine something like our gender, right? Right?

Radical Feminists

There are a number of good reasons to be glad you're not an insect: You have been spared the short life span, the squashability, the unappealing diet, the odd appearance, the general ickiness factor, and the difficulty of finding matching shoes for so many feet. If all that is not enough to persuade you of your astounding luck in having been born a mammal, here's another reason: *Wolbachia*.

Wolbachia is a whole family of bacteria that doesn't directly concern humans but is a huge health problem for insects. It has been estimated that at least one in every five insects on Earth is infected with some sort of *Wolbachia*.[4] Because there are a lot of insects out there, *Wolbachia* infection is probably the most common disease in the world.

Up until the 1970s, we weren't very interested in *Wolbachia*. Healing sick crickets isn't exactly on the top of anyone's priority list. The only reason for studying them would either be to find a way to use them to get rid of insect pests in agriculture, or to fulfil one's sheer curiosity.[5] But something has been discovered that's put a whole new spin on things: Different *Wolbachia* species have different tricks to en-

4. This includes spiders, worms, and other creepy creatures.

5. Or, of course, to be ready to fight when alien insects invade, in case those dumb sci-fi films turn out to be correct.

sure that they spread to as many hosts as possible, and they all involve taking control of their hosts' sex lives in a very female-favoring way.

Wolbachia are obligate parasites, which means they cannot live outside their hosts, and they can only spread vertically—insect offspring catch them from their parents before birth. *Wolbachia* usually prefer to be passed on from mothers to daughters, because female egg cells are a lot more hospitable than the tiny sperm cells of males. Here are some chilling examples of how they engineer this.

Some *Wolbachia* species will transform every male they infect into a female, through a process known as feminization. This is essentially a sex-change operation that's performed by manipulating the insect's hormones. Some male insects are even less fortunate: Their *Wolbachia* infection simply kills them off before birth—that way their sisters (also infected, of course) will have a good meal (McSibling!) waiting for them when they come to life, and the *Wolbachia* that they now house are more likely to thrive. Some *Wolbachia* are less lethal, but very fussy: If an infected male tries to mate with an uninfected female, there'll be no offspring. A successful match is allowed only between two infected partners. And, last but not least, males can be tossed aside completely: Some *Wolbachia* species encourage infected females to reproduce without males at all. This is called parthenogenesis or, less formally, virgin birth.

In the future, whenever you're having boy-girl type problems, reflect on all this, and thank your lucky stars that at least you've been spared this kind of confusion.

Bugs on the Brain

Madness. First it was the full moon, then it was demons. A hundred years ago, it was established that it was all the parents' fault. Then the blame shifted to bad genes. What next?

To understand what microbes may have to do with madness, we can start by going to the dogs. A rabid dog—a dog with the viral infection rabies—is a fearsome creature to behold, and a manifestly crazy one; at a certain stage of the illness, it becomes extremely irritable and aggressive, and tries to bite everyone in sight. Rabies is transmitted from one dog to another via saliva. When it finds its way into the bloodstream, it infiltrates the brain and settles there, multiplying quickly, before moving along the nerves to the salivary glands.

In Chapter 3, we saw how infectious microbes are able to manipulate their host's body in order to increase their own dispersal (coughing, sneezing, diarrhea . . .). The rabies virus is one of those parasites that has established a particularly insidious method of host manipulation—brain control. The aggressive behavior instigated by the rabies virus helps it move to further hosts: rabid dogs (and other canines and mammals) bite other dogs, spreading the disease, which travels via the biter's saliva into the bitee's body. Luckily, humans are among the hosts that don't transmit rabies to each other, but the fact that a human host is a dead end for the infecting virus doesn't help affected individuals—they will die if they've not been previously vaccinated or if they don't get treatment quickly.

Many other examples of parasites manipulating the behavior of their hosts are known throughout the natural world: Flatworms invade ant brains and make them climb onto grass stalks, where they are eaten by cows and sheep, instead of heading back to the anthill; the protozoan *Toxoplasma gondii* causes rats to lose their fear of cat smells; and one internal parasite makes small bugs jump into water, whereupon aquatic birds eat them. Precisely how these mechanisms work, or how they developed, is still an open question. The effects of the parasite on the host brain may be due to direct influence, or to the host's immune reaction to it, or both. They probably arose as an accidental side effect of the infection, and then became more pronounced as time went on.[6]

In principle, there is not much difference between one parasite exploiting its host's nervous system (brain included) and other parasites doing much the same thing to the respiratory, immune, or digestive systems. All these actions fall under the heading of what Richard Dawkins has titled the extended phenotype. Briefly, the effects of a gene are not restricted to the body in which that gene resides: A beaver dam is the result of the influence on the environment of beaver genes; a termite mound is the result of the action of termite genes; and a berserk dog is the result of the action of viral genes. No real difference here, except for the fact that the environment happens to be another organism.

The aspect that most disturbs us about all this, I suspect, is that our brain is what makes us *us*. I don't mind admitting that my runny nose is caused by a virus, but it would be very hard for me to accept that what I feel, and say, and do, and think may also be affected by anything of the sort. I'm me! I'm in control of my mind! There's no one else in here!

Nonetheless, evidence hints otherwise. It has long been known that some syphilis sufferers can undergo serious mental deterioration, as can people affected by Lyme disease, Chagas disease, and other infectious conditions (we have already seen what prions can do to a brain). In all these conditions, the dementia is caused by the

6. Nowhere is the influence of the parasite more evident than in the case of the emerald cockroach wasp, which lands on a cockroach and injects a cocktail of chemicals directly into a specific part of the roach's brain. This surgical strike causes the roach to lose any inclination to go anywhere—instead, it becomes content to let the wasp lead it calmly, by the antennae, to a secluded place. What happens to the cockroach next is rather disgusting, even by cockroach standards.

infectious agent destroying brain cells, left and right, but recent studies have been finding more subtle connections. Obsessive-compulsive disorder (OCD) and Tourette's syndrome have been statistically linked to prior streptococcal infections; *Toxoplasma gondii* (*T. gondii*), *Chlamydia* bacteria, and *cytomegalovirus* have been linked to schizophrenia; and other conditions, including autism and Alzheimer's disease, are also under investigation.

The effects of microbes may not even be confined to outright disease: Studies have linked *T. gondii* infections to lower intelligence, and also to thrill-seeking behavior. Because about 40 percent of all humans are infected with this parasite (including an estimated 50 million in the U.S.), this is a worrying sign. How much of our supposedly independent personalities and behaviors are affected by infections we don't even know about?

No one is suggesting that all, or even most, mental illnesses or personality traits are directly caused by microbial infection. The reasons for mental illness are diverse, and include genetic predisposition, developmental factors, parental upbringing, and a host of other parameters. The question of how infectious agents fit into this complex picture is a good one, and one that does not look likely to be answered in full anytime soon.

Although it's a scary thing to think about—it gives a whole new meaning to the term *infectious personality*—working out the role that microbes play in mental illness can be a good thing: Apart from the trivial observation that knowing the cause of a disease helps when it comes to treatment, the fact is that infections are sometimes treatable and, to an extent, reversible (something we can't yet say about genes or parents). Research into these issues needs to involve experts from diverse fields—from psychology, immunology, epidemiology, microbiology, and neurobiology. A natural tension exists, of course, between these various disciplines, with each one straining to exclude the others; but scientists can, if they want to, cooperate and inform each other's viewpoints. I hope they do. Any bugs lurking unnoticed in my brain probably don't share my enthusiasm, but I've still got the upper hand.

Or so you think.

Pardon?

Never mind. Forget we mentioned it.

Well, I sometimes wonder.

BONUS TRACK #8

Generations

I love my local library. It has a study area with a lovely verdant view—which is very good for inspiration—a water cooler, anda free wireless Internet access, (essential for my two primary activities: research and time wasting). Yes, my library's got it all. If it wasn't for the raucous Italian music blaring from the ground floor, generated by a lively accordion-guitar duo performing its heart out in the library as part of the local "multicultural week" program, this would have been a perfect place for writing.

A particularly vigorous rendition of "Papaveri e Papere" propelled me to the front desk to ask whether there was anywhere I could sit in relative quiet, whereupon the librarian bustled me into the "genealogy research center," a quiet little room with a few computers and a lot of books with a lot of names and lineages listed inside. Most important, it featured a door that closes. The librarian promised that somebody or other would probably come down to extricate me sooner or later, and she left. Ah, peace at last.

The records of past generations are particularly apt as surroundings for our next tale: It started in 1988, when biologist Richard Lenski took a single, unexceptional *E.coli* bacterium, put it in a flask filled with a pretty meager supply of nutrients, and bid it go forth and multiply. He separated its descendants into twelve populations and started growing them all in identical conditions. This has been going on since then. Each day for over twenty years the microbes have been transferred into new flasks with fresh nutrients, and every seventy-five days (five hundred generations in microbe time) samples are frozen for future study. These samples can be thawed and reexamined at any point, which is really nice because what you get by that is, to quote Lenski, a "frozen fossil record," a living archive that can show us exactly what happened and when it happened to whom—sort of a "rewind" function. If we want to, we can "replay" any part of it we like, letting, say, "generation 12,500" out of the freezer and growing it again and seeing if it evolves the same way twice. It also enables comparisons between generations, pitting "generation 25,000" against "generation 1,500" in a battle for control of a single flask, a feature that will be appreciated among sports fans who like debating whether the 1927 Yankees would've creamed the 1998 lineup, or how Ali at his prime would've matched up to Tyson.

This experiment is rather unusual: As a rule, when a scientist sets up an experiment, she'll have something specific she wants to measure, or an expected outcome she wishes to verify, a condition she wants to test. Questions like "What effect does temperature have on the development of the eggs of velocifrogs?" or "Can the chemical

Zapalot kill pancreatic cancer cells under conditions such-and-such?" and so on. This experiment, however, is open-ended: "Here are some initial conditions, now let's just carry on doing this for a while and see what happens." Not too many of these experiments are performed, on this scale at least, probably because few scientists have the patience (or time, or funds) to hang around for an indefinite period of time and hope for something interesting to turn up.

In this case, interesting stuff did turn up, although it took its sweet time coming: To nobody's surprise,[7] all the populations started evolving almost immediately. Some of the changes were pretty uniform in all the populations, and were clearly adaptations to the conditions they were under (although these similar changes often took different paths in different strains—a phenomenon known as convergent evolution). Some changes popped up in a few of the populations but not in others, which shows, as expected, that chance and randomness also have a role in evolution. So far, so good.

The real mind-blower came at around generation 33,127. It needs a bit of background: The only usable source of carbon these *E.coli* were given was glucose. The flasks also contained citrate (the tangy-tasting acid that gives lemons their zip) that *E.coli* can't use. But suddenly, one flask showed the cloudiness that suggested citrate was being broken down. Outside contamination? No. A strain of *E.coli* had somehow acquired the ability to metabolize citrate.

It's not easy to explain just how shocking this development is. Citrate digestion is not a simple skill for a microbe. I can roughly compare it to a human being evolving a way to digest grass. After 33,000 generations of steady glucose munching, one microbial cell was suddenly endowed with this special ability and exclaimed, "Hey! This other stuff's food, too!" and, aided by its enhanced nutrition, proceeded to outbreed the hell out of all the rest of the *E.coli* in the vicinity, leading very quickly to its being the dominant type in any flask it was placed in. Textbook evolution. What's more, as I mentioned before, Lenski and his colleagues can go back through their "fossil record" fridges and find exactly what sort of mutations happened, when, and in what order. They could also "rerun" the events to see if the same things happen again. They can do all sorts of stuff. And they don't need to dig up fossils or reconstruct bones or carbon-date; they can just revive the "ancestors" and analyze them. This is great. If only human genealogy

continues

---

7. Nobody except creationists, of course. But then creationists don't like this experiment *at all*—you'll soon see why—and have been avoiding any discussion of it like the plague.

were this flexible. Instead of this book-filled room I'm sitting in, there would be a warehouse full of caskets, and every time somebody wanted to know how come he's the only redhead in the family, or what *really* happened to great-great-great-great grandfather Ebenezer back in 1875, he could just unlock a casket and . . . hmm, perhaps not such a good idea after all. Let's stick to discussing microbe lineages, at least until I leave this room.

It's worth mentioning here that when we test for *E.coli,* in water samples or food or whatever, one of the routine tests is for citrate metabolism. If it metabolizes citrate, it isn't *E.coli.* Now a bacterium has come forth that is obviously *E.coli,* and yet it does metabolize citrate. Either we redefine our categories, or we have to say that this is a new sort of microbe. If that's the case, then we have just witnessed the formation of a new species.

The terms *species* and *microbes* aren't easily reconciled. As I've mentioned before, what exactly constitutes a different species in microbes is an open question, one that depends more on our definitions of *species* than on the microbes themselves. What is very clear is that a new *function,* a new capacity, has sprung up from, well, essentially from nowhere. Now *that,* my friends, is evolution.

This experiment and other experiments in a similar vein are showing microbes evolving all kinds of things, from various, sometimes unexpected, adaptations to the environment; to ruthless competition for nutrients; to differentiation, cooperation, and even mini-ecosystems springing up. New worlds are being created inside flasks, and we watch the coming generations as they unfold and bloom before our eyes.

And speaking of generations, I think I'll go home to my son now.

EPILOGUE

Q: So . . .

A: Thanks for reading this far. I'm very glad you made the time.

Q: Are you done?

A: Hardly. There's so much more I wanted to squeeze in; some of my favorites ended up on the cutting-room floor, for one reason or another. Maybe someday I'll do a director's-cut edition with magnetic bacteria, and Carl Woese's revolution, and the ambergris commercial. Also, I feel that fungi have been grossly underrepresented here, which is a rotten, moldy shame. Marvelous creatures. Perhaps next time.

Q: Quite the microbe fanatic, aren't you?

A: Not really, no.

Q: Huh?

A: Well, microbiology is just one way of understanding what's going on around us. There are plenty of other places to begin that journey. I like microbes because they offer a fairly simple, basic route to tread. But if you're curious about life and the universe, you could start off just about anywhere.

Q: And where would we end up?

A: That's for each of us to see for ourselves. Personally, I don't like ending up anywhere at all. You've probably noticed that an irritatingly large number of the sections in this book end with unanswered questions, disputes, and conflicting views. That's because interesting stuff keeps happening at the edges, where not much is resolved yet. Science is a journey, not a destination. Anyone who presents science primarily as an authoritative, well-organized body of established facts is probably trying to sell you something.

Q: So what you're saying is that science is just one alternative out of many for truth seekers?

A: No, I still think it's special. Not perfect, not the only one, but special. The scientific method has the concepts of change and objectivity deep inside its basic structure. They help balance our very human tendency to see the world as we want it to appear. Science also appears to be very useful for keeping planes up in the air, keeping people healthy—that sort of thing.

Q: And yet the book is filled with examples in which scientists were wrong.

A: Being wrong comes with the job. Scientists investigate uncharted seas of knowledge and experience, so blundering about is a very common occurrence. Besides, scientists are every bit as human as anyone else: They play politics, and they get stubborn, and defensive, and greedy, and territorial—and, on occasion, just plain dumb. What makes science different is that, at one point or another, an error can be found and corrected. Disagreements are ultimately resolved, not carried on ad infinitum. That's the intention, at least.

Q: Is the information in this book guaranteed 100 percent correct and up to date?

A: By the time you read it, almost certainly not. One detail or another has changed since I last looked. Things move so quickly in biology that keeping up with everything is nearly impossible, especially for someone as slothlike as I am. I can assure you that I have tried very hard to ensure that only the highest-quality, freshest facts and theories available were used in the making of this book.

Q: You appear to be a very dirty-minded person, constantly hammering on about sex, and disease, and death, and excrement.

A: Yes, that's true.

Q: Final question. Does the book have a point?

A: There seems to be some sort of message about our place in nature buzzing around the text. What it is exactly I'm not sure; I'm still figuring it out myself. But the main thing I wanted to show was how fluid and varied life is—how it always manages to surprise us. Science (like baseball, or anything else) is only boring if you don't understand what's happening. Reality's deliciously absurd and fun—dig in.

ACKNOWLEDGMENTS

My first and foremost vote of thanks goes to the hundreds of scientists, present and past, whose discoveries provided the material for this book. In the lab, the clinic, the sea, at the desk, up volcanoes, and down mines, researchers rock.

This book owes its existence to the Melbourne University Writing Centre for Scholars and Researchers, and its director Simon Clews, who provided me with extralarge helpings of support, encouragement, and instruction—and never once batted an eyelid at my method of scheduling discussions, which involved popping into his office unannounced and yapping away. Simon was instrumental in introducing me to my splendid literary agent, Clare Forster, who nurtured the book from its disjointed beginnings, and guided its bewildered author through the intricacies of the publishing trade with good humor, unflagging patience, and impeccable professionalism.

I am deeply indebted to Professor Christina Cheers of the University of Melbourne for going over the first draft of the manuscript, making many invaluable comments, and preventing me from making a fool of myself. Michael Brand also provided many illuminating remarks. Grating use of language in an early version of the text was cheerfully weeded out by Eleanor Heft. Julian Frost thought this book was a good idea, even before I did, and helped make it so, in more ways than one. Thanks to Miki and Eitan Shapiro for the stimulating conversation, and to Trish O'Connor for her good cheer and helpful suggestions.

Not last and certainly not least, my indomitable editor at Scribe, Nicola Shafer, who spent her last days as a carefree bachelor(ette) ensuring that my sentences eventually came to an end at some point, that terms such as *directed mutagenesis* were translated into Human, and that the book did not degenerate into a monstrous, serpentine, many-headed digression. This American edition is the direct result of the enthusiasm and professionalism of the Basic Books editorial gang—Amanda Moon, Michelle Welsh-Horst, Eleanor Duncan, and Whitney Casser—displayed in the face of a tight schedule and an emphatically overseas and occasionally obstinate author.

Despite the sterling efforts of all the abovementioned, this book undoubtedly contains some errors, for which I, and I alone, am to be held responsible.

My former supervisor at the University of Jerusalem, Professor David Yogev, through his verve and enthusiasm for microbiology, was a source of inspiration to me.

I am apologetically grateful to Professor Glenn Browning and Dr. Phil Markham at the University of Melbourne, who displayed saintly patience and kindness towards me, with scant returns. Likewise, Professors Paul Griffiths and Ofer Gal of the University of Sydney were exceedingly understanding during all the times that I gave my attention to writing this book instead of writing my PhD.

Finally, my thanks to my outrageously supportive family, in-laws included; and to my wife Tamar, for whom words fail me.

GLOSSARY

aerotolerant microbe: an anaerobic microbe not inhibited by the presence of atmospheric oxygen (O_2).

allele: an alternative form of a gene; one of the different forms of a gene that can exist at a single locus (spot on a chromosome).

amino acids: large organic molecules that constitute the basic building block of proteins. There are twenty common amino acids that link together, in various orders, to form proteins in animals. The order of amino acids is determined by the animal's genetic sequence.

anaerobic microbe: a microorganism that grows in the absence of atmospheric oxygen (O_2).

annamox: anaerobic ammonium oxidation; a biological process by which ammonium (NH_4+) and nitrite (NO_2-) are converted directly into gaseous nitrogen (N_2).

antibiotic: a chemical substance that kills or suppresses the growth of microorganisms.

antigenic variation: the process by which an infectious organism alters its surface proteins in order to evade a host immune response.

antioxidant: a substance that inhibits oxidation and is capable of counteracting the damaging effects of oxidation in body tissue.

archaea: a microscopic, single-celled, prokaryotic form of life that forms one of the three domains in the tree of life. The other two domains are bacteria and eukarya.

asymptomatic carrier: a person or animal who is infected with an infectious disease but who displays no discernible symptoms. Although unaffected by the disease or the disorder themselves, carriers can transmit it to others.

attenuation: reduction in the virulence of a virus, either as a natural process or artificially for the purpose of creating a vaccine.

bacteria: microscopic, single-celled organisms; all prokaryotes not belonging to the domain archaea.

bacteriophage: a virus that infects bacteria.

bacteriovorus bacteria: predatory bacteria that feed on other types of bacteria.

bacterium: a single bacterial organism; the singular form of *bacteria.*

bioremediation: the use of microorganisms or plants to remove or detoxify toxic, or unwanted, chemicals in an environment.

cell: the fundamental unit of life; the simplest unit that can exist, grow, and reproduce independently.

cellular membrane: the membrane that separates the cell from its outer environment, and that regulates the flow of material into and out of the cell. It is composed mainly of lipids, with embedded proteins contributing important regulatory functions.

cellulose: the main carbohydrate in living plants. Cellulose forms the skeletal structure of the plant cell wall.

chemotaxis: movement by a cell or organism towards or away from a chemical.

chromosome: a package of wound-up DNA in the nucleus of a cell containing the cell's genetic material. Prokaryotes typically have a single, circular chromosome, while eukaryotes typically have several chromosomes (twenty-three pairs in humans), each containing a linear DNA molecule.

conjugation: a microbial sexual process by which two single-celled organisms exchange a portion of their DNA across a bridge formed between the cells.

cryptobiosis: a state in which an animal's metabolic activities come to a reversible standstill.

denitrifying bacteria: bacteria capable of reducing nitrite (NO_2-) to gaseous nitrogen (N_2), under anaerobic conditions.

DNA: deoxyribonucleic acid; the genetic material of all living cells and some viruses.

elasticotaxis: the ability of an organism to sense and respond to elastic forces in the surface supporting it.

endosymbiosis: an organism that lives within the body or cells of another organism. Eukaryotic cell organelles, such as mitochondria and chloroplasts, are believed to have originated as free-living bacteria that were then incorporated into early eukaryotic cells.

enzyme: a protein molecule that acts as a catalyst in mediating and speeding a specific chemical reaction.

epidemiology: the study of the incidence, prevalence, distribution, and control of a disease in populations.

epithelial cell: a cell that covers a surface in the body such as the skin or the inner lining of the digestive tract.

eukaryote: an organism or cell that has its DNA in a clearly defined nucleus and usually has other organelles, too. Compare with a prokaryote.

evolution: the change in the genetic makeup of a population of organisms over time.

facultative anaerobe: a microorganism that is able to grow in either the presence or absence of atmospheric oxygen (O_2).

FDA: U.S. Food and Drug Administration; regulatory body for the development, approval, manufacture, sale, and use of drugs in the United States.

fermentative microbe: a microorganism carrying out fermentation, the process of energy production in a cell in an anaerobic environment (with no oxygen present).

flagellum: a long, whiplike tail protruding from the surface of a cell that propels the cell, acting as a locomotive device.

free radicals: short-lived, highly reactive molecules that have one or more unpaired electrons. Their high reactivity can cause damage to living cells.

fungi: a kingdom of eukaryotic organisms (an independent group equal in rank to that of plants and animals). Fungi cannot carry out photosynthesis and are more closely related to animals than to plants. Types of fungi include molds, yeasts, and mushrooms.

gene: a unit of heredity; a segment of genetic material (typically DNA) that specifies the structure of a protein or an RNA molecule.

genetic code: the nucleotide sequence of a DNA molecule (or, in certain viruses, of an RNA molecule), which contains information for the synthesis of proteins. It's used by the cells to translate genetic information in a protein sequence: Three successive nucleotides in a gene form one codon, and each codon calls for a single amino acid.

genetic sequence: the precise order of bases in a nucleic acid.

genome: all the genetic information encoded in the DNA of a living organism's cell.

horizontal gene transfer (HGT): the process in which an organism transfers genetic material to another cell that is not its offspring.

hormone: a chemical substance secreted in one part of an organism and transported to another part of that organism, where it has a specific effect.

hyperthermophilic bacteria: an organism that can live at very high temperatures. A hypothermophile's typical optimum growth temperature is above 176° F.

inflammation: a body's protective response to injury or infection.

lysogenic cycle: integration of a bacteriophage nucleic acid into the host bacterium's genome. The integrated genetic material can then be passed on to the host bacterium's offspring, until favorable conditions cause its release and the start of the lytic cycle.

lytic cycle: the main strategy of a virus infection. Consists of the virus penetrating the host cell, taking over cellular machinery to produce viral components (which it then assembles into active viral form), and lysing (bursting) the host cell, before emerging to infect further host cells.

magnetotaxis: the ability of an organism or cell to sense a magnetic field and coordinate movement in response to it.

messenger RNA (mRNA): an RNA molecule transcribed from DNA that contains the genetic information necessary to encode a particular protein.

metabolism: the sum of all physical and chemical changes that take place within an organism, and all energy transformations that occur within living cells.

methanogenesis: the production of methane gas as a metabolic by-product.

methanogens: archaea that carry out methanogenesis.

microbe: a microorganism; any organism that is too small to be seen clearly with the naked eye.

mitochondria: membrane-enclosed cellular compartments that are the major source of an eukaryotic cell's energy.

molecular biology: a branch of biological science devoted to the study of the structure and function of biological molecules, with particular attention to DNA, RNA, and proteins.

molecule: two or more atoms held together by chemical bonds.

mutation: an inheritable change in the base sequence of the genome of an organism.

National Institutes of Health (NIH): the principal biomedical research agency of the U.S. federal government.

nitrogen fixing: the conversion by certain soil microorganisms of gaseous nitrogen (N_2) into compounds that plants and other organisms can assimilate.

noncoding DNA: stretches of a DNA sequence that do not code for known genes.

parthenogenesis: a form of reproduction in which the egg develops into a new individual without fertilization by sperm.

photosynthesis: the use of light energy to drive the incorporation of carbon dioxide (CO_2) into cell material.

phototaxis: the movement of an organism or cell in response to light.

pilus: a filamentous appendage that extends from the surface of the bacterial cell. Used in bacterial conjugation.

plasmid: a circular DNA molecule that is not part of the chromosome and that replicates independently of it.

prion: an infectious particle that does not contain DNA or RNA, but consists of only a hydrophobic protein; believed to be the smallest infectious particle.

prokaryote: an organism lacking a nucleus and other membrane-bounded compartments. Bacteria and archaea are prokaryotes. Compare with eukaryote.

promoter: the site on a DNA sequence where the RNA polymerase binds and begins transcription.

protein: a molecule composed of amino acids linked together in a particular order. Proteins perform a wide variety of functions, including serving as enzymes, structural components, or signalling molecules.

protists: single-celled eukaryotic organisms.

recalcitrant: a substance that is difficult to degrade under natural conditions; a substance resistant to microbial attack.

receptor proteins: a protein molecule that enables a cell to recognize a particular chemical.

revertant: a mutant that has reverted to its former genotype or to the original phenotype.

ribosome: the cellular structure where messenger RNA (mRNA) is translated into protein.

RNA: ribonucleic acid; a molecule similar in structure to DNA that plays important roles in protein synthesis and other chemical activities of the cell. RNA is the genetic material of some viruses.

RNA interference: a mechanism, recently discovered, for RNA-guided regulation of gene expression. Small RNA molecules (referred to as small interference RNA or siRNA) can induce sequence-specific silencing of gene expression.

RNA polymerase: an enzyme that makes an RNA copy of a DNA or RNA template. In all cells, RNA polymerase is needed for constructing RNA chains from DNA genes (a process called transcription).

spore: a general term for resistant resting structures of microbes. In fungi, the spore is the basic reproductive body; in prokaryotes, it is a dormant form that a cell develops to protect itself for extended periods of time in adverse conditions.

strict aerobe: organism requiring atmospheric oxygen (O_2) for survival.

structural proteins: proteins located in or on the cellular membrane that interact with the cell's outside environment.

substrate: the molecule undergoing reaction with an enzyme; any substance an enzyme acts on.

transcription: the process by which DNA is used as a template by the enzyme RNA polymerase for the synthesis of an RNA molecule; the first step in protein production.

transduction: the transfer of bacterial genetic material between bacteria by a bacteriophage.

transformation: transfer of genetic material via free DNA; uptake and incorporation of exogenous DNA into a cell.

translation: the process by which the genetic code carried by mRNA directs the production of proteins from amino acids. Takes place in the ribosome.

transport protein: a protein that transports a molecule within a cell or within a biological fluid.

transposon: a DNA element that has the ability to move around to different positions within the genome of a single cell in a process called transposition. Often also carries genes unrelated to transposition.

vertical gene transfer (VGT): the transfer of genetic material to an organism through ancestral inheritance.

virus: microscopic infectious agent (many of which are pathogenic) that replicates itself only within cells of living hosts; a piece of nucleic acid (DNA or RNA) wrapped in a protein coat.

FURTHER READING

Dozens and dozens of research and review articles, books, magazines, Internet sites, and other sources provided the material for this book. I've decided to save a tree or two by not listing them all. Instead, I'll provide a small selection of good places to carry on reading about the subjects I've been discussing throughout this book:

BOOKS

Angiers, N., *The Canon* (New York: Houghton Mifflin, 2007).

- An enlightening, crystal-clear general-audience science book.

Bryson, B., *A Short History of Nearly Everything* (New York: Broadway, 2003).

- A general-audience science book that needs my endorsement like Eric Clapton needs my guitar-playing tips.

Crawford, D.H., *Deadly Companions* (New York: Oxford University Press, 2007).

- A good book that tells the story of bad epidemics.

Dawkins, R., *The Blind Watchmaker* (New York: W.W. Norton & Company, 1986); *The Selfish Gene*, 3rd rev. ed. (New York: Oxford University Press, 2006).

- Two powerful, eye-opening introductions to evolutionary thinking.

Dawkins, R., *The Extended Phenotype* (Oxford: Oxford University Press, 1982).

- A more technical work from Dawkins.

Dembski, W. and Ruse, M.A. (eds.), *Debating Design* (Cambridge: Cambridge University Press, 2004).

- An infuriatingly balanced account of the intelligent-design debate, with contributions by major players on both sides.

Dixon, B., *Power Unseen* (Oxford: Oxford University Press, 1994).

- A good place to go to for some more microbes.

Drexler, M., *Secret Agents* (Washington, D.C.: Joseph Henry Press, 2002).

- More on pandemics, for people who fall asleep too easily.

Madigan, M.T., Martinko, J.M., Dunlap, P.V., and Clark, D.P., *Brock Biology of Microorganisms*, 12th ed. (San Francisco: Benjamin Cummings, 2008).

- The definitive microbiology textbook. It's comprehensive, readable, and an invaluable resource for any microbe enthusiast. I was raised on the knees of the ninth edition.

Margulis, L. and Sagan, D., *Microcosmos* (New York: HarperCollins, 1987); *What is Life?* (Berkeley: University of California Press, 1995); *Dazzle Gradually* (Vermont: Chelsea Green, 2007).

- Margulis, who formulated modern endosymbiotic theory, provides a good and necessary balance to Dawkins on evolutionary theory.

Moalem, S. (with Price, J.), *Survival of the Sickest* (New York: HarperCollins, 2006).

- Medicine, genetics, and evolution all fit together in surprising ways.

Ridley, M., *Genome* (New York: HarperCollins, 1999).

- A grand tour of the human genome.

Sterelny, K. and Griffiths, P.E., *Sex and Death: An Introduction to Philosophy of Biology* (Chicago: University of Chicago Press, 1999).

- A good source to turn to if you feel that my more abstract wonderings and digressions left something to be desired.

Waller, J., *Fabulous Science* (Oxford: Oxford University Press, 2002).

- Revisits several scientific events and careers with exhilarating, iconoclastic results.

Watson, J.D., Baker, T.A., Bell, S.P., Gann, A., Levine, M., and Losick, R., *Molecular Biology of the Gene*, 6th ed. (San Francisco: Benjamin Cummings, 2008).

- The one-stop-shop for all your gene-related needs.

Zimmer, C., *Microcosm* (New York: Pantheon, 2008).

- Everything you wanted to know about *E. coli*—and much more, in this new Zimmer book.

Zimmer, C., *Parasite Rex* (New York: Free Press, 2001).

- An enchanting account of all things parasite.

SCHOLARLY ARTICLE

I will mention just one particularly enjoyable scholarly article: "Size Doesn't Matter: Towards a More Inclusive Philosophy of Biology" by Maureen A. O'Malley and John Dupré, in issue 2, volume 22 of the journal *Philosophy and Biology*. Accessible at http://www.springerlink.com/content/122x56horv566h67/

JOURNALS AND MAGAZINES

Nature

- The foremost science journal, aimed primarily at the scientifically literate.

Nature Reviews Microbiology

- My favorite professional source for microbiological information.

New Scientist

- British popular-science magazine. The place to go for science news and articles.

Science

- Another leading science journal.

Scientific American

- *New Scientist*'s American counterpart.

BLOGS AND WEBSITES

Carl Zimmer's blog, The Loom, http://scienceblogs.com/loom/

- You should definitely check this out.

John Wilkins's blog, Evolving Thoughts, http://scienceblogs.com/evolvingthoughts/

- Another lovely evolution-related blog.

Microbeworld, http://www.microbeworld.org/

- Site including photos and podcasts and links.

Moselio Schaechter's blog, Small Things Considered, http://schaechter.asmblog.org/

- A wonderful repository of microbial info, discussions, and links. I just wish I'd discovered it a bit earlier—it would have saved me a whole lot of research effort.

Tardrigrades, http://www.tardigrades.com/

- An unpretentious site that is the place to go and see just how cute the water bears look.

Finally, should you wish to know more about anything mentioned here, you are most cordially invited to drop me a line at idan.invisiblekingdom@gmail.com. See you there.